A GREEN ARMAGEDDON

A Green Armageddon

Tom Brown

The Book Guild Ltd.
Sussex, England

The Book Guild Ltd.
25 High Street
Lewes, Sussex

First Published 1993

Set in Baskerville
Typesetting by R & H Typesetters
Horley, Surrey
Printed in Great Britain by
Antony Rowe Ltd.
Chippenham, Wiltshire.

A catalogue record for this book is
available from the British Library

ISBN 0 86332 795 8

CONTENTS

ACKNOWLEDGEMENTS

To my wife, Alice, who impatiently checked and typed the manuscript and to colleagues, friends and acquaintances who in the past may have inadvertently stimulated some of my ideas and in so doing contributed to this book.

Thanks are also due to Michael Dewhurst of Soil Fertility Planning for his advice and constructive criticism.

APOLOGIES

I have only dealt with certain aspects of our environment about which I have some knowledge. There are many more features which have not been mentioned. One of these concerns that most populous and popular animal — the fish. For this I apologise but suggest there are more capable hands than mine to bridge the gap. I do remember some thirty to forty years ago reading a book entitled *The Overfishing Problem.* History and events in our environment, it seems, have a habit of repetition.

INTRODUCTION

> *'There are two things I am very confident I can do very well: one is an introduction to any literary work, stating what it is to contain, and how it should be executed in the most perfect manner . . .'*
> (Samuel Johnson)

I make no claim to have Johnson's ability, but merely state that an author and a book have to begin somewhere, and an introduction is as good as any other means for starting a book. The first thing that people might say if they ever read the title of the book or note the name of the author is: 'Who is Tom Brown and what does he know about the environment?' I would answer, very little, but there is a wealth of environmental reading material available to enable the most uniformed layman to acquire a smattering of intelligence about it, and if, like me, you are a proverbial doubter and are capable of constructing a simple sentence, what more justification need there be for writing it down on paper? The stimulus to write what follows came in the early summer of 1989 whilst scything thistles in Home Paddock. Although on such an exercise the arms are fully engaged wielding the implement, the mind is unoccupied and free to range . . .

What I hope I have done is to present various aspects of our surroundings in a context which is readable, avoiding the pedestrianism of the textbook.

I

THE GREEN REVOLUTION

'When you destroy a blade of grass
You poison England at her roots.'
(Gordon Bottomley)

Each morning as we emerge from our beds to face the daily traumas, we must first cope with the news media which regularly brings to our attention world events considered to be essential spice to our breakfast cereal — wars and rumours of wars, democracies waxing and waning, terrorism and Watergateism, famine, desolation and the daily obituaries. The 1980s were not untypical of such a cocktail, but historians may remember and cite this particular period as that time when the world gave birth to a new kind of religion, the hallowed name of which has become a veritable hip-word to be heard on the lips of down-and-outs and a new concept amongst politicians in the halls of power — ENVIRONMENT. This eleven-letter word is a conditioned belief, and rather like the Church of England appears to have always been with us, but people have been so busy trying to make ends meet, that little attention has been given to the condition.

An early definition of the environment was given by Aristotle when he suggested that the elements consisted of earth, air, fire and water, items which of course are still with us. How do we relate to this 'thing' which envelops us, is there in the morning when we wake up and is there waiting whilst we are fast asleep? It's been there since before Adam's time and although Methuselah is reported to have lived the best part of a thousand years we are not told whether he was aware of the environment.

It is not such a static quantity as would appear but changes slowly over millennia. Some people, who may or may not be alarmists, claim that humans have accelerated this slow rate of change — only history will tell how successful or disastrous this has been. None of us, however, can deny that prediction of Daniel that 'men shall run to and fro' as wishful thinking, although he didn't quite visualise how. The pace is nevertheless increasing and suggests an unfulfilled human yearning for the instancy of magic as when the water was turned into wine. The real problem is that the other constituents in the environment cannot keep up with this pace and many fall by the wayside.

The more astute historians will know that the authentic birth of environment occurred in the form of a government department in the 1970s to provide jobs for well-meaning but unimpressive politicians whose main raison d'être was to stop would-be entrepreneurs from building unsightly edifices in important people's back yards. It was in the 1980s, however, when a group of people searching for political recognition, suddenly fell upon the idea of calling themselves the GREEN PARTY, and for a cause célèbre, they adopted the environment.

The first seedlings of the Green Party had been around in the sixties and appeared as unemployed, bearded and sandelled prophets accompanied by their bespectacled tweed-skirted socialist soul-mates delivering messages that were either quickly forgotten or totally ignored. They had a preference for vegetarianism, opting for brown rice and over-spiced greens and were often assessed by more sensitive people as being somewhat economical with soap and water. They preached against violence, supporting the liberalisation of sexual offerings. Since there were no conflagrations of note going on at the time, they divided their attention between demonstrating against the folly of nuclear war and waving anti-hunt banners.

But seedlings grow and, like any other biological entity,

they are subject to conditioning and evolution. What we are witnessing today, not just locally but world-wide, is the flowering of an idea which although having very inauspicious beginnings, is now attracting serious attention. This success has been achieved partly because of a less than attractive characteristic of human nature — an inclination to disrupt. It is often evident whenever children are gathered together, such children being described as difficult to control or advanced for their age. It is a habit less vulgarly apparent in adulthood, but there are those among us in whom it persists as a means of attracting attention. These are the people who constantly search for the faults in society, expose them, exaggerate them, institute a cause to alarm the general public of the dangers they are facing and then rally the support of people in the public eye who know little or nothing of the problem in hand. This unfortunate habit is called lobbying and is a common deliberation in parliamentary procedure. It is well illustrated by the Greens (a familiarity this party favours) actually rallying the support of the Prime Minister. Mrs Thatcher in her inimitable manner did not admit to *supporting* Green policies or even being influenced by them. It was all a very far-seeing idea of hers in the first place, and no doubt she would like to be remembered by posterity as the first Green Prime Minister.

From time to time during the formulation of Green Party policies what was happening in the world either suddenly became prominent and was noticed, or became fundamentally changed. An example of the latter is vegetarianism. The diets of many Green members which hitherto had been taken for granted suddenly acquired the importance which questioned how it arrived on the table, in particular the meat portion. It was suddenly realised that for years meat consumption had involved killing and killing was cruel. There is now a strong leaning among Greens towards a green diet.

Happenings in the world which suddenly became prominent resulted from the advantages of increasing wealth. More and more visitors to Los Angeles were appalled by the smog and feared for the rest of the developed world. The air we breathe no longer consists of that original Garden of

Eden mixture of one-fifth oxygen and four-fifths nitrogen. The energy of ancient history was discovered to inhabit channels deep beneath the earth's surface and once harnessed has begun slowly but with increasing regularity to change the composition of the atmosphere.

It all started when someone, somewhere, some time ago (about 3000 BC) discovered that a circle accelerates the pace of life. Having no beginning and no ending it was worshipped as a symbol of perpetuity and Fortune. The circle gave birth to the wheel, and the wheel gave birth to the motor car which in the twentieth century is worshipped as a sex symbol and has to be constantly satiated with injections of historic energy. Cars have now become bigger, faster, gallon-guzzling polluters and whilst the horse-and-cart age is looked back on through rose-tinted glasses by organic gardeners among the Greens, most of them, running fast cars, have no desire to return to stabling. In some future age when the fossils have been depleted there may be a short respite of fresh air pending the development of some new means of acceleration.

But to move on. The media, particularly television, has helped to emphasise the contrast between the overfed, oversexed developed world where the pleasures involved in sexual engagements are not inconvenienced by any resulting encumbrances, allowing populations to remain static, and the Third World where the populations are underfed, still oversexed but without restrictions and increasing at alarming rates. The Second World never comes up for discussion but must be a half-way house.

The difference between the financial security and well-being of the prince and that of the pauper has been exhaustively observed, annotated and satirised and appears to be an on-going condition. In order to compensate the pauper for his unenviable status, religions throughout the world have promised him better things in the next life. Christianity has however found some difficulty in deciding the fate of the wealthy. 'Solomon in all his glory' was never short of a penny or a woman but was never specifically indicted. Presumably there will be special dispensation for the more prominent among the well-endowed, exempting them from that mind-bending problem of negotiating the eye of a needle.

But promises for the future which at best are only assumptions do not satisfy continually hungry mouths, nor are their wretched owners inspired when they hear talk of surpluses, economic strategy, set-aside and monetary compensation amounts — a foreign language to them, but the staff of life to European bureacrats. Most of us in the West are immune to the meaning of famine, perhaps because of over-exposure by the media, and can readily ease any slight discomfort from time to time by dropping our small change in the poor-box.

The humanity which has developed the complex techniques to tackle space appears to be consistently incapable of coming to terms with that so much more mundane responsibility of producing and distributing enough food for all its members. The Darwinian primary concept of self-preservation may well be of paramount importance to the individual, but in collective terms it becomes divisive to the extent that some individuals are more worth preserving than others. The Darwinian secondary concept of self-reproduction is now *the* dominating factor and will surely get completely out of hand unless something akin to Aids or the condom can halt this progression. Is it then just the stuff of fiction to wonder whether some future being, engaged in classification of extinct species, will append a note that the age of humanity was shorter than that of the dinosaurs because the hominid, Homo Sapiens, did not possess sufficient intelligence to control its own numbers?

Once the Green ball started to roll there was no stopping it. Acres of materials began queueing up for recycling. Every movement, every system, every mouthful now has to be meticulously monitored to prevent mankind's slow drift to extinction becoming an avalanche.

Global warming and the greenhouse effect are esoteric terms which have been discussed for many years among scientists but have now transferred directly from clinical science laboratories into beer-stained taprooms. These kinds of phenomena, whether real or imagined, have a cataclysmic, Star-Trek fascination for the viewing public who have already digested the far greater threat of world population increase but continually need to keep changing the diet to maintain the excitement. In the final analysis there is an economic limit beyond which the wealthiest nations cannot afford to go to offset what is after all only an estimated atmospheric Armageddon, and there are those behind coming slowly up the track who understandably are keen to enjoy the fruits of development before the end draws nigh.

A much more down-to-earth problem taken up by the Greens, and appearing to be insoluble, has emerged as a result of government policy to keep roadside verges well trimmed. During the summer months, fleets of orange tractors are to be seen mowing these verges, and there is then revealed the evidence of how wealthy the nation has become — able to throw away so much indestructible material. The well-heeled members of society travelling in the fast-lane never notice this accumulation of detritus. It's the poor sods on the inside lane, driving twenty-year-old bangers who have suddenly become aware of the throw-away complex. Even this awareness has been a rather slow process because most of the stuff was deposited there by them in the first place. But the message has come through loud and clear. One only has to see the car in front flick out a sweet paper in transit — the headlights are violently switched on, the klaxon horn is depressed and the offending car is overtaken at break-neck speed, the guilty litter-bug at the same time receiving the benefit of a purple-faced glare from a Green-conscious member of the public.

Clearly the Greens are infiltrating every corner of the globe, expressing a wide range of opinions on a library of subjects from ozone gaps in space to empty cigarette packets in the gutter, from poisons in our food and waterways to canine toiletry in our city parks. The pragmatists and the environmentalists in our midst are lining up to do battle for the truth. Whilst there are some people who can be fooled all the time, the rest will claim they are never fooled. But even if we are to believe that destiny is inescapable and cannot be denied, the truth nevertheless may never be fully disclosed or understood. Are we witnessing the inevitable wheel-grind of an accelerating evolution or is it all a pseudo-scientific and political ploy to catch the eye of governments and improve the status of the Green Party?

2

FOOD AND THE FUTURE

'Blessings on him who invented . . . the food that appeases hunger . . .
(Miguel de Cervantes)

Food provides the essential energy by which we all survive — in truth we eat to live. It did not take humans very long, however, to become rather bored with the very basic kind of nutritional fare on offer and along with so-called civilisation (a unique kind of evolution conceived by the hominids to distinguish them from less arrogant species), the human diet began to reflect more the pleasurable experience of living to eat. The human diet has now become sufficiently sophisticated to tempt the over-pampered consumer rather than to provide a necessity. To achieve this end, an elaborate stratification from ground to gullet has been developed partly to provide useful employment for those not engaged directly in food production, but mainly to seduce the consumer.

This phase is now on the wane, because the developed world has suddenly become aware that living to eat is not good for human metabolism. In the eighteenth century, Brillat-Savarin, a French gastronomist, claimed that 'the destiny of the nations depends on what they eat.' The need not only to temper the quantity but question the quality of the intake is paramount to promote good health (a condition taken for granted by those who survived in primitive tribes). A re-affirmation of this essential precept was made when the Green Party came on the scene and in a very short time they turned the agricultural world upside down in an effort to change our way of life, but in particular our diet.

The reason for this sudden change of emphasis is not readily identified, but it may have been given a boost because of people's reluctance to pay taxes. The West became too efficient in providing the dietary needs of its peoples and surpluses resulted. Magna-granaries of grain, magna-cold stores of beef and lamb, lakes of wine — it all seemed to be a reflection of some gargantuan squirrel complex — and it had to be paid for and then stored at enormous cost. Consumers suddently realised that they were the ones who were paying for all this profligacy and the more Green-minded among them began to ask questions.

After the 1939-45 war, agricultural production began to move upwards. Detailed attention had been paid to ways and means of achieving this during the war years in order to ensure that the nation was well fed. When all the restrictions were over, the swords were beaten into ploughshares and everything associated with agriculture was regarded as a bull market with no limits. Research went into overdrive to produce multiple blades of grass where only one grew before, aided and abetted by the multinationals producing feeds, fertilisers and agro-chemicals. There was a bland complacency that growth was going to be permanently exponential. Any suggestions that there might be a production ceiling beyond which it would be imprudent to expand was never considered, or if it was, never discussed.

At that time I was working with some very bright researchers, but I do not recall the word surplus ever being mentioned. The recent research cuts may be a drastic form of nemesis but that drive to produce more and more and more had a Gadarene complex about it and some such outcome should have been anticipated in the seventies and the research emphasis changed. What is sad is that there is so much more important work to be done in agricultural research than that just concerned with increasing production, but governments, like banks, work on a relatively short-term financial basis and in the case of agricultural research were readily able to find adequate reasons for bringing in the axe.

The wheel has come full-circle and the aim now is to reduce these embarrassing surpluses by various methods of lowering production. The Greens have been prominent in supporting these moves and at the same time have been giving attention to the constituents of our diet to expose to the world that not only has agriculture gone awry quantitatively, but the quality is not there. Our diet is indeed full of toxins.

Mention has already been made that exaggeration is a way of life adopted by pressure groups to attract attention. In any pressure group, there are elements which whilst endowed with only fragile intelligence are sufficiently vociferous to find an audience. The group as a whole is not harmed by such mindless clamouring because publicity of any kind is good for business and groups like the Green Party have certainly made their mark by way of exaggerated claims and accusations. The margin between these various allegations and downright untruths is rather hazy because there is usually an element of truth in what is said. The questions that were asked in the seventies when surpluses began to mount were concerned with the quality as well as the amount of food produced. In short it would appear that we are all being systematically poisoned by the farming fraternity in order to keep farmers in businss.

When members of the public read that 'modern farming methods contaminate the farm with food poisoning organisms', that 'so much is being done to destroy our planet', that 'farm animals and crops are brimming with toxic residues', then they sit up and take notice and react in a predictable manner which is to look for a scapegoat before examining the evidence. The finger is pointed at the farmer who represents a convenient source for all the ills that men are heir to. Quite large sections of the public who would claim to be literate have succumbed to the hypnotic repetition of such alarm-bell ringing words as toxicity, fungicides, herbicides, pesticides and drugs — all potential killers if they manage to infiltrate into the frail human body.

When these gypsies' warnings were combined with that world-wide warning of an approaching atmospheric apocalypse, the Green Party succeeded in securing 15% of the European vote, little short of miraculous for a first attempt. Apart from their extreme views, the Greens have brought a much more instant awareness of what goes on around us, but what is more to the point, they have impressed what have traditionally been unimpressionable — governments. There always have been 'things' wrong with our environment and no doubt there always will be. It seems a pity that, to draw attention to what needs to be done, strategies are employed which depend upon questionable shock tactics. We have for years accused communist states of brainwashing their peoples into beliefs which we discarded at the end of our student days, but are still prepared to listen and respond to people who preach half truths and are expert in innuendo. If you are one of the gullible public, take notice of this prophecy from the High Seat of Parliament. 'The effects [greenhouse] may be horrific. Europe could expect famine and food shortages. Coastal areas could flood and freshwater shortages with climate warming would hit crop yields with certainty. All this could happen within the next sixty to a hundred years.'

The remarkable thing about it all is that every Joe Soap down the street is talking about it. My own reaction is to feel sorry for those poor sandwich-board men who, for a pittance, year in year out have marched up and down the seaside promenades, strolled slowly past lines of hopeful voyeurs queueing for the X-film, mixed with the townies at agricultural shows, carrying fore and aft their prophetic messages: 'Flee from the wrath to come,' 'Believe on the Lord Jesus Christ and thou shalt be saved,' 'Repent ye, for the kingdom of heaven is at hand,' — but barely a worthless soul took any notice. I may not be here in sixty to a hundred years' time, but I have a gut feeling that this disastrous prognosis is too hypothetical. I would like to say I stand to be corrected, but by the time there is any real evidence one way or the other, this would be a physical impossibility. If I had to choose between what the 'informed' regard as a firm prophecy and what can only amount to a guess — that man will muddle through somehow — I would opt for the latter.

In this year of our Lord 1990 on the 10th day of December when the snow is thick on the ground in that New Jersalem of England's green and pleasant land, I read on page one of an old lady sitting crouched over a one-bar fire clothed in her entire wardrobe to keep out the cold. Then I turn to page two and read the velvet words of a politician telling me how much the Welfare State is providing for the poor and needy. On page three the scientist tells me what I should do in preparation for Global Warming and I then begin to wonder whether economy with the truth and sensationalism are not just the faults of the dying twentieth century but are essential tools in the power struggle as we approach its last decade and prepare for the coming-of-age of century twenty-one.

3

UNSAVOURY SALTS

'. . . but if the salt have lost his savour, wherewith shall it be salted?'

(Matthew 5:3)

The easiest target for the Green Party has been nitrates in our water supplies. Most people have little idea what nitrates are except that they are salts and have something to do with nitrogen. Even the experts cannot agree on what has become known as the nitrate problem. Critics claim that most of the nitrogen applied to the land in the form of fertiliser ends up polluting our rivers, seas and atmosphere rather than helping plants to grow. Part of the atmospheric pollution stems, indirectly presumably, from too many cudding animals (and perhaps an increase in vegetarianism amongst the Greens).

The problem starts with a supposition, not from laymen but from trained scientists, which states that nitrates above a certain level are poisonous. Research institutes have been conned into persuading governments to limit nitrate inputs internationally, and governments in turn have been passing these instructions on to farmers.

For those who have forgotten or never learned about nitrogen and nitrates the information which follows may help.

1. Nitrogen is a constituent of protein and is

essential to plant growth.

2. The atmosphere contains four-fifths nitrogen but most of this is not available to plants. The exceptions are the legumes (clover, beans, peas, lupins, laburnum etc.) which are able to 'fix' atmospheric nitrogen and store it in their root nodules in the form of nitrate ready for subsequent use.

3. Whilst animals and humans need nitrogen, they cannot assimilate directly but obtain it via plants (the human also via animals including fish). The nitrogen contained in the food eaten by animals and humans is then returned to the atmosphere after biological action including denitrification to produce what is generally known as the Nitrogen Cycle.

4. The soil contains nitrogen in organic form, not available to plants until it has been 'mineralised' by bacterial action into an absorbable inorganic form. This process is slow and for most farming purposes there is insufficient of the resulting nitrate to support adequate plant growth.

5. Farmyard manure, slurry and compost are 'natural' products from the farm where animals are kept — compost need not necessarily have ingredients originating from animals. These materials have varying degrees of ability to produce nitrate which essentially is in organic form and has to be broken down by bacterial action to produce inorganic nitrates. These products are bulky and expensive in terms of storage, transport and spreading. They must be utilised at some stage since they cannot be stored indefinitely.

6. There are natural deposits of Chilean nitrate which are in limited supply and expensive.

7. Inorganic nitrate can be manufactured. Nitrogen is extracted from the atmosphere using North Sea gas for energy. There is also a by-product of the steel industry called ammonium sulphate which was a regular fertiliser used during the war years.

8. Nitrate fertilisers are not generally applied on their own but in conjunction with potash and phosphate, a common collective term for the three being NPK.

The nitrate problem originates from the policy established soon after 1939 which invested farming with the essential role of feeding a nation under virtual siege. Production of foodstuffs, to replace imports, had to be increased and the manner of achieving this was by the increased use of ammonium sulphate relative to other fertilisers (apart from farmyard manure), largely basic slag for phosphate. The resulting boom in production particularly after the war centred on the increased use of nitrogen fertiliser. Since it was cheap, the tendency was to apply it in excess to ensure adequate plant growth. For the last thirty years farm production has increased by about 3% a year but little thought was given to what happened to the surplus nitrate not taken up by the plant.

What would appear to be a straightforward leaching of excess nitrate into our water system is far from the truth, particularly if we realise that it may take twenty to thirty years for such leaching to reach underground aquifers; what is more, nitrate leaching can occur without fertiliser application.

What really initiated the problem was when some bright scientist discovered that in the laboratory, nitrates reacted with proteins to form nitrites and further reacted to produce nitrosamenes which are carcinogens. The word cancer has only to be whispered along corridors occupied by pressure groups and there is a frantic scrambling for the panic button. In comprehensive studies of stomach cancer, there was found to be no correlation whatsoever with high nitrate levels in drinking water. The Greens, nevertheless, had a field-day particularly when nitrates were linked with a condition in babies called Blue Baby Syndrome, on very tenuous evidence.

It is significant that as nitrate levels in water have increased, this sydrome has all but disappeared. Nothwithstanding the lack of evidence, the panic button had been pressed and the World Health Organisation, which is naturally very Green, together with the European Community, dreamed up a figure of 50 milligrams of nitrate per litre of drinking water as the maximum amount allowable, and this stricture will become compulsory in the EC from 1992 onwards. One must assume that any concentration over this level will begin to have dire consequences on public health, though no-one as yet has been bold enough to say precisely what and how. The cost of carrying out this directive wherever it is applicable — usually good arable land or first-class pasture — is likely to be enormous.

To put this problem in perspective, some recent work carried out at Shinfield with pigs, animals which have a very similar metabolism to that of the human, showed that when offered feed containing nitrates 100 times greater than the maximum recommended level of 50mg/litre, the pigs remained perfectly healthy. The nitrates were fed in milk, and one is perhaps led to ask, are there any other sources of nitrate apart from water which can be taken in by the consumer? The answer is that 80-90% of nitrate uptake is via vegetables. Clearly the Greens and especially the vegetarians among them never addressed themselves to this question. Vegetables do in fact contain far higher levels of nitrate than the EC limit and the nitrate content can deteriorate with excess storage to form N — nitroso compounds which are anything but health-promoting. The vegetarian Greens have been so concerned with all the other poisons alleged to be residual in the farm produce that the regular intake of nitrates from their dinner plates was never given a thought. If it had, they would have realised that the idea of increased nitrate intake leading to cancer is a non-starter. Vegetarians, who it has been established have a higher nitrate intake than meat-eaters, do exhibit a lower incidence of cancer than the average. The assumption might be that meat is the cancer culprit, but such an assumption would be far too simplistic.

The outcome of all this nitrate nonsense is that initially farmers will be expected to adhere to the stricture voluntarily but inevitably there will be a compulsory requirement. Compensation levels are being worked out but most of these nitrogen-sensitive areas (NSA) are on good fertile land and many farmers are likely to be put out of business. The USA appear to be leaders in the paranoic reaction to nitrates and their restriction is even lower than that of the EC.

The general consensus of opinion is that high nitrogen input via artificial fertilisers is not the only culprit. Both farmyard manure and slurry are also implicated. In research at Rothamsted, untreated soil has been showing a steady nitrate content for years. Halving the input of nitrate does not necessarily halve the leaching. One contributory factor to leaching is ploughing up old pasture. During the 1939-45 war this procedure was taken to the extreme to include golf courses. All the nitrogen that had been locked up in these pastures suddenly became available via mineralisation and may well have contributed to some of the nitrate levels in water that we are finding today. Nitrate levels in water at that time were not considered as a threat to life and limb.

Various bits of advice have come from the Ministry.

1. No fertiliser application in the autumn or heavy application of FYM or slurry [the normal time for spreading].
2. Don't leave ploughed-up pastures bare, i.e. without a crop to utilise all the mineralised nitrate.
3. Don't use excess amounts of nitrate in the spring.
4. Modify stocking rates and use more clover in sowing grass seeds mixtures to replace 'bagged' nitrogen.

There is a break point beyond which nitrate application of any kind is undesirable economically or environmentally, but this point could change from year to year, from season to season, and from area to area locally depending on weather, soil type, stocking rates and farm-programme. Any attempt to monitor these variable reactions and arrive at a general conclusion is not going to be very rewarding. Perhaps the only acceptable generalisation is that nitrate leaching is more likely to occur on grazed grassland than arable farms (where the total crop is removed in one season), and that the less favoured areas in the uplands are less likely to be restricted than the highly productive lowlands even though some watercourses in the uplands have high nitrate levels. Wherever there is moderate to high rainfall, nitrate concentration is likely to be diluted. Thus the West Country will have lower concentrations than the Eastern counties.

If all goes according to European Community plans the South East corner of England and hundreds of hectares in Nitrogen Sensitive Areas will be turned into grassland prairies, national parks, golf courses, recreation areas for townies and 'natural' habitats for biologists. A start has already been made, and the comforting thought is that when the panic is over and the Community bureaucrats become convinced that nitrates are no longer lethal, there will be a very respectable amount of nitrogen locked up in the organic safe beneath the sod waiting to be mineralised — but at what cost!

4

THE NEGATIVE ELEMENT

'Cure the disease and kill the patient'.
(Francis Bacon)

The environment contains what might be regarded as positive elements which promote good health and negative elements which induce the reverse — disease. For a given genetic make-up probably the most significant factor for maintaining good health in plants and animals of any description is nutrition. As yet there is no satisfactory control of genetic-borne diseases, but the problems they give rise to can be minimised by treatment and the appropriate nutrition. The dictum that prevention is better than cure is regularly trotted out to emphasise what the right approach should be, but knowing precisely what to prevent would be more meaningful. Disease often 'arrives' before there has been any anticipation of it, largely because of a lack of awareness, or to put it less euphemistically because of ignorance. A typical instance of this was the 'arrival' of bovine spongiform encephalitis (BSE) which caused cattle to assume a mad gait, stopped people eating beef, and left both government and researchers with red faces.

The origin of BSE appears to be related to nutrition or, to be more precise, the constituents of the feed. For many years farmers have fed products of animal origin to pigs and poultry which are omnivorous. At some point a decision was made to feed similar products to cattle and sheep which are herbivorous — a course of action which might be described as a delayed form of cannibalism. The rendering process to provide this fare was evidently adequate to destroy any potential disease organisms which might have been transmitted. There was no consequent disease outbreak until the rendering process was changed — presumably to one that was inadequate.

The symptoms and lesions of BSE are similar to those found in sheep infected with scrapie, a disease of the central nervous system. The incidence of clinical scrapie in sheep is low and it is likely that the disease in cattle originated from the rendered products of sheep which carried a sub-clinical infection. All this trouble started when the intention was taken to provide cheaper protein. Presumably the compounders were advised by the scientists on the safety of the process and the 'appropriateness' of the resulting feed. The government too should have been aware of what was going on. The ineptness of both bodies was shown when following the outbreak, an announcement was made to the effect that progeny of BSE-affected dams was unlikely to be infected. There was no sound scientific evidence whatsoever for this claim. The government exacerbated the problem by offering farmers only a percentage of the real value of the infected animals, allowing diseased stock to slip through the net.

At no time during the early development of the saga was any thought given to public safety. This was left to the media who had a field-day. British beef was declared unfit to eat at home and abroad, without any evidence, and the public reaction, cutting beef out of the diet, served as a warning to both government and industry that if you play with fire, you are likely to get burned. The responsibility for this chain of events lies with the compounders who suffered no real loss but merely changed the feed constituents. What appeared to be most revealing was that farmers were feeding this infected material quite unaware of its origin because there was no obligation on the compounders to disclose the feed ingredients.

Although it was the dairy herds in which the disease was primarily diagnosed, beef producers suffered when consumption of beef dropped dramatically. It may have been only a nine-days wonder for the press, but the knock-on effect on all meat consumption was felt throughout the livestock section of farming.

The BSE scare followed on the heels of what can only be regarded as a fiasco illustrating the readiness with which government officials rush towards Dad's Army-style panic stations. A combination of pseudo-scientific overstatement and a lust for publicity resulted in thousands of healthy hens being destroyed, farmers put out of business, and a squawking politician with a bird brain the size of a hen's egg being dropped from the government. The media has only to spell out the word SALMONELLA, and the whole country is immediately under a state of siege. Millions of eggs are consumed daily and although there were only a relatively small number of cases of salmonellosis, the government took those drastic killing measures and then compounded the idiocy by replacing the egg deficit by unmonitored imports.

There is always the outside chance that causal organisms producing diseases like BSE or salmonellosis will disappear or become extinct, but until they do, we must assume that they are a relatively permanent feature of the environment, and the positive strategy is to try and minimise their presence or effects. All organisms at every level possess the means to resist conditions which detract from their welfare. Thus aphids have the ability to excrete substances called pheronomes which sensitise or warn other aphids of imminent dangers (parasites or sprays). The higher animals are endowed with a blood system which can harness antibodies when challenged by a disease infection. Disease organisms have this 'will' to live which, whilst quite unconscious, nevertheless enables them to face adversity. Thousands of tons of herbicides and pesticides have been pumped out of tankers throughout the world since the 1940s, but the bugs are still with us. The virus causing the common cold is still active and so are the rabbits however much their population was ravaged by myxomatosis. It is regularly claimed that warble-fly in cattle and sheep scab could be eradicated if simple measures were comprehensively followed. Can these bugs numb human decisions? Rabies, eradicated in this country, is said to be waiting in attendance just across the Channel hard by the entrance to the Eurotunnel. It is in fact far easier to kill the patient than cure the disease.

The drive towards greater intensity of production since the last war has meant that bugs of all kinds have proliferated. To cope with this onslaught, agro-chemical companies have taken root to the extent that there are available nowadays at any time of the year some 4,000 herbicides, 3,000 fungicides and 700 pesticides and as the bugs become resistant, the lists change. Whilst most of these compound chemicals are specific to the bugs concerned, there are residues left behind in the crop which, although infinitesimal, nevertheless could affect some consumers, particularly people with allergies. The safeguards against this happening are most stringent — the testing time for any new product is four to five years and there is currently a ten-year waiting list. The cost of developing such a product is in the order of £15 million and there is then a regular re-appraisal at ten year intervals.

The advent of Green Party policies has directed public attention into what goes on on the farm. The impression has been given that we are all in grave danger from the products of the agro-chemical companies. It is, of course, difficult for these companies to refute such implications without being accused of bias. They do have vested interests and they have invited criticism by not having given sufficient attention to alternative treatments. There are some moves in this direction, but the progress is slow and the criticism remains.

The farmer, who is a layman with regard to the constituents of these agro-chemicals, is becoming increasingly bewildered by the range of sprays and drugs he is recommended to adminster to his crops or his animals. It is understandable that he is beginning to look around for alternative means of treatment, under pressure from welfare and Green groups and the public in general, to stop 'poisoning' his land and his animals, however misinformed these allegations might be. There is always this rumour of long-term effects likely to remain unproven except by hindsight. When we take aspirin from time to time we can never be absolutely sure there are no long-term effects. There is in fact little absolute certainty about anything but there is a tendency to generalise from the particular, to leap from isolated incidences of toxic residues to the assumption that all produce provides a risk. Dipping sheep using organo-phosphorus dips and spraying greenhouse crops with insecticides expose the operators to the risk of breathing in minute particles of poison. If precautions are not taken or are inadequate, there could be long-term cumulative effects on the operators but not on consumers of lamb or greenhouse crops.

Paracelsus in the 1400s aptly defined a poison: 'Everything is a poison, nothing is a poison, it's the dose that makes the poison.' Whenever we eat or drink there is a reasonable chance that we are taking in some poisons, but at such a low concentration that no ill-effects follow. To ensure that these levels remain low, government regulations stipulate the maximum permissible concentration of various poisons. Cyanide and arsenic which have featured in many real-life murders and fictional whodunnits are permitted in concentrations of fifty parts per billion. Herbicides and pesticides of any sort must be no more than 0.1 parts per billion, i.e. 500 times less and barely measurable. Such safety measures, whilst extreme, do take into account the possibility of accidents, errors or stupidity on the part of the operator. Testing laboratories too are not totally immune from the occasional error. Where proper care is taken, following precisely the operating instructions, there should be no danger to the operator.

Nevertheless the customer is always right, and the demand now is for consumer goods which have not been treated with 'killer' sprays. The consumer is prepared to pay for such goods and it behoves both research bodies and agro-chemical companies to invest in a future which offers a more 'natural' means of pest control.

The range of treatment for livestock, whilst not as varied as that for crops, is still substantial and so is the cost as stock numbers have increased. Chemical fertilisers released the potential of grass growth and as the numbers of livestock increased so did the parasites, which was good news for the agro-chemical companies.

The most persistent parasites affecting livestock out at grass are the roundworms and means to control them have reached the position where whole groups rather than single species can be knocked out. Since these parasites spend a part of their life-cycle outside the host on the grass, it is conceivable that this would be a convenient opportunity to eliminate them. No satisfactory treatment, however, has yet been devised to achieve this.

The knock-out process mentioned above is not a once and for all treatment. There is a wealth of infection in reserve but, more seriously, resistance to drugs has become a way of life for the roundworms, so much so that reports are coming in from Australia, New Zealand and South Africa, countries where sheep are kept in vast numbers, that this resistance is considerable. A biological entity of any kind, we have already suggested, is inordinately difficult, if not impossible, to eliminate by drugs and any such treatment of the host animal needs to be combined with the kind of management which seeks to break or interfere with the life-cycle of the parasite. Freedom from parasites can only be achieved successfully indoors under aseptic conditions, and as far as herbivores are concerned is of no practical value. Biologically the hosts, be they poultry, pigs, cattle or sheep, need to be challenged by the parasite in order to stimulate those antibodies which provide resistance. It is when this challenge goes over the top, as in intensive systems, that the host cannot cope, the resistance breaks down, and symptoms of sub-clinical (weight loss) or clinical disease are exhibited.

New methods of improving or increasing production are constantly being researched and presented to the farmer as Utopias. Such discoveries do not automatically have public opinion or public health in mind. One of these discoveries is bovine somatotropin (BST).

The general trend in dairying has been to reduce expensive inputs like concentrates in order to moderate the flow of milk. Too much butter and cheese have been going into European food banks to be stored at taxpayers' expense. The idea of stimulating even more milk production would seem to be a nonsense — except perhaps to allow a reduction in the herd numbers.

BST is a natural hormone already possessed by the cow, but by injecting extra amounts into the cow's bloodstream, the mammary glands can be stimulated to produce more milk provided extra feed is available. Now whatever the argument for or against, the customer is always going to object to milk 'full of hormones'. This is a misinterpretation of the effect and there is no foreseen danger from such milk, but the Green Party will no doubt insist that, if such milk comes on the market, it should be labelled, and quite rightly so.

What is a justifiable objection to this promotion is that the throughput of feed has to be increased in order to allow the stimulus to function adequately. Has the cow the physical capacity to take in the extra bulk, or are we back to the already rejected idea of feeding more concentrates?

A more serious objection from the consumer has been in relation to growth promoters. In the 1970s it was found that antibiotics appeared to stimulate liveweight increase by reducing the number of pathogenic bacteria in the gut, allowing the beneficial ones to proliferate. Since that time a whole new range of growth promoters has been developed including steroids.

Where residues of some growth promoters appear in the dung which is then spread on to pasture, the animals subsequently grazing the pasture may be at risk. Similarly any residues which appear in meat may be a risk to the public. There is no valid reason to use such growth promoters. The simplest and most natural way to promote growth in animals is by a balanced diet, supported by a good environment which does not impose excess strain, and allows the animals freedom to forage.

The latest research is that concerned with bio-technology. It aims to change by genetic means the type of animal or plant we require. Thus genetically identical cattle, presumably with built-in resistance to some of the common diseases, and pest-resistant plants are likely to be somewhere in the pipeline. This type of technology will replace embryo-splitting on the animal side and pest-resistant plants could well supersede agro-chemicals, but pests and diseases being what they are will find their own means to survive. When Adam and Eve were turned out of the Garden to face the thorns and thistles of the outside world they could not have contemplated the battle that was to ensue between the positive and negative elements in the environment and all because that fateful apple was eaten.

5

VEGETARIANISM AND SPECIALISM

VEGETARIANISM

'He's content with a vegetable love
which would certainly not suit me.'
(W. S. Gilbert)

The dentition of Homo Sapiens is equipped to negotiate both vegetables and meat, but unlike the herbivore, which is multi-stomached, man has but a 'short' stomach, not designed to deal specifically with an exclusively vegetarian diet. Primitive man did however manage to survive largely on 'greens' and roots spiced with beetles and various comatose insects which could be readily grabbed. Small rodents and fish were a relative luxury, and anything larger a positive banquet.

In the twentieth century (and of course even earlier) meat has replaced vegetables as the main constituent of the diet and is valued nutritionally because it is a complete food, i.e. it contains all the essential vitamins to promote good health. For many years the 'meat and two veg' menu constituted the rather unimaginative offering of the average British household and less salubrious restaurants. But with the passage of time and cultural progress, food has gradually become more civilised. The steady increase in world trade has meant that many vegetables once regarded as exotic in this country are now readily available, a boon to consumers who prefer little or no meat, providing them with choice and variety in a diet which hitherto had been largely monotonous and boring.

Before the last war in this country, boiled cabbage was a staple ingredient of the diet to complement the beef, potatoes and Yorkshire pudding on Sundays. It was overcooked, of doubtful nutritional value, but left its mark in the atomosphere, particularly in those households where the cooking facilities were invariably part of the living-room furniture. The smell of stale cabbage was often enriched by the somewhat different odour of 'bubble and squeak', the cold cabbage and potato remains left over from Sunday fried up on Monday. Only those fortunate ones who relaxed in the luxury of separate kitchens and indoor toilets were aware of this smell whenever they had to cross the thresholds of the poor on goodwill missions.

In the early days of Christian belief, the vegetarian cult emerged for religious reasons, in complete obedience to the instruction: 'Thou shalt not kill.' Killing is not a pleasant job at any time, depending, of course, on what you are killing. Most people don't turn a hair when killing a fly or setting a trap for the house mouse. Killing larger sentient animals is quite a different matter, and the more hypersensitive members of society find this quite obnoxious and have become vegetarians. But rather in the same vein as when missionaries safaried to Africa to convert the natives to Christianity, vegetarians would have us all follow their example. Fortunately some kind of balance is present in nature whereby we are not all of one persuasion, and even, though such a motley set of beliefs has led to conflicts of various intensity, there are still animals about in the fields to ensure a more satisfactory and savoury survival for mankind.

Christian and other religions which advocate vegetarianism do so however strictly they may adhere to the teaching and doctrine of the Bible. At the time the Mosaic Law was allegedly written, animals had little protection from any well-meaning groups and were regularly sacrificed for one reason or another. The Old Testament is replete with instances of killing lambs, calves and goats to appease Jehovah's wrath, indeed the preference for meat rather than 'greens' could be said to have incited the first Biblical murder. Cain's offering of 'fruit and veg' found little favour with the Almighty who preferred instead the bloody sacrifice of Abel. Only Cain lived to tell the tale. All this bloody slaughter may well have persuaded some religions to opt for vegetarianism.

In America, religious belief goes hand in hand with respectable wealth, well illustrated by the Kellogg family who expounded the vegetarian cult. The modern cornflake, however, has probably done more for the entrepreneurial pocket than it has done for the religious conviction of the soul.

Whilst meat sales in developed countries, particularly in America, have boomed, the consumers of meat have not necessarily benefitted by enjoying good health. In any programme sampling good health amongst the population, vegetarians usually fare comparatively well, and it is this fact plus the slide towards obesity among some of the older generation which has persuaded many youngsters in the twentieth century to become vegetarian.

There are some people around, not completely convinced or committed, who are prepared to take the risk of including 'white' meat in their diet. They blandly choose their menus under the misconception that white meat is safe, red meat is not. The white meat choice is confined to poultry; the pig is red as far as they are concerned. There is some psychological influence here which persuades them that red spells danger whereas white has a purity about it. They are completely unaware of the number of additives that go into the production of their white meat compared with the virtually nil amount used to produce red meat in the UK.

There are other people similarly selective who will eat fish but not meat. The preparation of fish for the table does not involve the traumatic association of the slaughterhouse and the spilling of blood. There is no sound of objection as with pigs squealing, sheep baaing and cattle mooing — all the preparation is in fact done on board ship away from sensitive eyes. Catching fish is regarded as a sport with very little opposition from the public, but fish do have nervous systems and will react to pain by struggling for example when caught on the fish-hook. In general terms fish have been subjected to less interference from human activity than farm animals (apart, that is, from farmed fish) and in that sense provide a 'purer' source of protein. There are people in the world whose diet is almost exclusively fish who would have great difficulty in changing their diet to one exclusively vegetarian.

Vegetarians also believe that their kind of diet poses less danger to health. In the sixties groundnut cake imported from Brazil was found to contain a fungal toxin (in the kernel). Many turkeys and ducks died after being offered feed containing this material. Groundnuts or peanuts are the ingredient for peanut butter, a favourite spread for vegetarians, particularly children. A similar toxin was found in samples of maize in India in the seventies. Clearly the origin and processing of vegetarian products need to be monitored just as much as those of meat.

An extreme form of vegetarianism is exhibited by vegans. These are individuals who are obsessed with the fear or worship of animals in the same way that chain-smokers are addicted to tobacco. Not only do they not eat meat but they abhor the feeling of being touched by any material derived from an animal, preferring instead to hide their bodies inside man-made materials which reach the shop counters via the profligate use of fossil fuels. For this reason they were not to be seen walking the streets before the twentieth century for fear of incarceration.

A life without meat and fish is conceivable. A life without milk and eggs and wool could be a life without animals. This is barely conceivable and vegans do not appear to have any rational ideas which might be of value to vegetarians.

One of the main reasons for the current interest in vegetarianism is the cruelty factor. There is accusation, with some truth, that cruelty is involved in the rearing, transporting and slaughtering of farm animals. The dictionary definition of cruelty may throw some light on the question. 'Cruel' is defined as either inflicting pain without pity, or causing pain or suffering. The latter definition could be used to describe the pain suffered when subjected to medical or dental attention but is not commonly ascribed to cruelty. A legal definition is to cause danger to life or limb, or a threat to bodily and mental health. These definitions suggest that cruelty can mean the infliction of pain with or without pity, deliberately or otherwise. They do not indicate whether there is satisfaction or pleasure involved.

If we examine the relationship between animals in their natural environment, clearly there is both pleasure and satisfaction. A cat will 'enjoy' playing with a mouse to the extent that once the mouse becomes comatose, but not necessarily dead, the cat may walk away. There is no further resistance or escape attempt to provide stimulus. The inverted commas around the word 'enjoy' indicate the human interpretation of the cat's behaviour, a form of anthropomorphism which will be referred to later. The lioness, stimulated by hunger, gains satisfaction from stalking and killing her prey, a necessary adjunct to survival. Killer whales are reported to 'sport' with dolphins and seals before killing them. Humans get some satisfaction reducing the fly or vermin population on their premises if only to avoid the possible incidence of infection or disease.

People have quite different feelings about animals and, apart from the various allergies, have their preferences and dislikes. Familiarity may increase or decrease affection towards animals. The genuine dog breeder or animal artist will develop an increasing affinity with dogs or all animals. The man responsible for killing sheep at the slaughterhouse is unlikely to have this kind of affinity with sheep but may well be a dog lover. How do we differentiate between the 'toff' paying to kill birds on the grouse moor, and the slaughterhouse man who is paid to fulfil an essential role in the food chain?

Hunting, particularly stag hunting, is a very contentious pursuit, so much so that some people, who would wish to be considered non-violent, behave quite differently when they start to wave anti-hunt banners. Opposition of any kind involves some sort of violence, initiated by verbal abuse, and as more emphasis is required becoming physical. Most country people supporting hunting would claim to be animal lovers, but when a joint master of stag hunting threatens to shoot all his hounds if a ban on stag hunting is imposed, he can have little or no affinity with animals, electing to use them as a bloody demonstration to show his support for hunting. Stag hunting is a country pursuit and those who support it are not necessarily cruel. It is one way of culling the deer population and allegedly of maintaining the herd, although both claims are arguable. Townspeople unfamiliar with a country way of life consider hunting of any kind barbarous and on a par with bull fighting, i.e. merely providing satisfaction for primitive instincts. Culling baby seals lying helpless on the beaches is also regarded as barbaric, and although the culling is necessary, it is the method which is unacceptable.

One of the underlying reasons for all this opposition is the attractiveness of the animals threatened. We are in fact indulging again in the kind of anthropomorphism which theorises that any animals that give us pleasure should be protected. The rat in the sewer provides no visual pleasure and would receive no votes for protection. The size of the animal too is important, the larger the animal the greater likelihood that it will be protected. The stag receives more support than the fox, the elephant than the pigmy shrew. Shakespeare recognised that the difference between the large and the small animal is only one of degree. In *Measure for Measure* Claudio states:

> '. . . the poor beetle, that we tread upon,
> In corporal sufference finds a pang as great
> As when a giant dies.'

The wild rabbit, about the same size as a hedgehog, is undoubtedly attractive but unlike the hedgehog is a persistent pest. Few people, therefore, complained about the disfigurement of the rabbit from myxomatosis.

The purposes intended by 'nature' do not necessarily correspond with how humans respond to them. The colour and song of birds, whilst attractive to the human eye and ear, have the sole purpose of either establishing territory or enhancing display as a sexual attraction. The pleasure that humans gain from this habit is purely incidental. Protection of the peacock and the song-thrush is therefore likely to be far more popular than protection of that uninspiring ubiquitous bird — the sparrow. When we do take sides, are our judgements sufficiently sound to decide whether the one is right and the other wrong?

The most contentious of all subjects regarding human/animal relationships is that of vivisection, the definition of which is experiments on animals which involve cutting into or dissecting the body. Such operations may be carried out whilst the animal is alive, but comatose, or may terminate the experiment and, at the same time, the animal's life.

Paracelsus in the fifteenth century was one of the first physicians to carry out vivisection, but it was probably a declaration by Descartes, the father of modern philosophy, which really propelled the medical profession into regarding vivisection as an essential study. He claimed that animals feel no pain because they are not conscious (like man) of being. That this claim could be believed illustrated the gullibility of medical hopefuls who considered themselves in intellectual ascendancy. They had only to look back in history to read how plebeians had enjoyed tormenting animals, but more pointedly to look around them at a time when bear-baiting and cockfigting were common pastimes, and dogs were bred and trained to rip to pieces in public any animals that the owners decided would provide 'sport'.

Some of the more outrageous bloodsports are now banned, but a relationship has now developed which purports to use animals in any way considered appropriate to promote the health and well-being of humans.

In 1965, Brigid Brophy wrote these words in the *Sunday Times*: 'We employ their work, we eat and wear them . . . Whereas we used to sacrifice them to our gods . . . we now sacrifice them to science . . . in the hope or on the mere off-chance, that we might see a little more clearly into the present.' What has changed in this country by and large is the deliberate infliction of pain on animals to provide pleasure to an audience; any cruelty involved in vivisection is incidental and not intentional. It is likely that people working in this type of medium have become immune to any feeling that what they are doing might be regarded as objectionable.

In 1975 a picture appeared in the *Sunday People* of beagles being forced to smoke in order to assess toxicity of tobacco.

I was appalled for two reasons, first, that dogs could be used for such an unworthy cause and secondly because the person in charge was a pupil in my school class in the 1930s — an exceptionally bright lad, but what a waste of talent.

The argument for vivisection is the furtherance of medical knowledge in the interests of human society. Among all the animals tested only the pig has a metabolism comparable to that of humans, so that tests on animals in general are not necessarily applicable to humans. A prime example of this was the test for thalidomide which produced no apparent reaction on animals, but the end effect on humans was horrendous.

An example in reverse is Fleming's penicillin. Had it been tried on guinea-pigs, it would never have seen the light of day. Penicillin is a poison as far as guinea-pigs are concerned.

Many tests on animals have quite a frivolous objective — for example to ensure that ingredients used in the cosmetic trade are not harmful — equivalent to trapping animals to provide furs for the luxury market. Not so very long ago, these activities were deemed to be perfectly satisfactory, but attitudes change. Television has brought into our homes a whole new biology and at the same time a completely revised concept of animals and plants. What goes on behind closed doors can no longer be kept secret, and the media make it their business to expose any activity which might be regarded as unacceptable by the reader, the listener or the viewer.

Worldwide, there is now considerable objection to vivisection because much more appropriate and acceptable tests can be carried out using organ and tissue culture. In the UK it is claimed that some three million animals die each year in the cause of research. Some of these will lead quite contented lives and be well fed and cared for. Some will undergo tests which may involve pain. In perspective the Royal Society for Prevention of Cruelty to Animals has to put down well over 100,000 animals each year because people no longer want them. This is no passport to their use for vivisection, but it does indicate that whilst cruelty may be banned officially, in circumstances where it is not deliberate, it is still practised by members of the public deliberately. Cruelty by human on humans has been common for generations and there is no evidence of any significant change. Cruelty to animals can only be controlled, it cannot be eliminated, so long as it remains part of the human character.

There are a number of questions to be asked in human/animal relationships which are not readily answered to everyone's satisfaction. A popular conception is one of the right to live. Does this assumption have any real meaning without qualification? In Genesis we are told that Adam was given life, but he may well have become bored without companionship. We have a right to live free from the various conditions which threaten our contentment, but then this right is often absent. The quality of life indeed is of far more consequence than a life which amounts to mere existence. Without contentment this right to live becomes rather meaningless and there is an arguable case for euthanasia where life becomes unbearable.

As far as animals are concerned, has the fly any right to live; has the rat or locust or any sentient animal a right to live? How do we differentiate between sentient animals and the bacteria and viruses which threaten our well-being? A plant, maltreated, will wither and die. It may be grown for our pleasure or our surivial, but it has not the equipment to complain. Does it have a right to live? Quite recently the common adder has been put on the short-list of rare species to be protected — never mind how poisonous the animal is, if it's in danger of extinction it's protected, so long as it poses no threat to humans. The same sort of assessments are made in the antique world — anything that is in short supply (throw-away Victoriana) is of value. Is there any rational sense in such evaluation?

Any animal that provides visual pleasure is deemed to need protection if this is judged necessary for survival until or unless that animal poses a threat to the well-being or survival of humanity. Should we differentiate on this basis or is it a case that man, because of arrogance, must take priority and decide whether the one should survive and the other die? Careful scrutiny of questions of this nature suggest that they are more readily posed than answered.

Most vegetarians and vegans do not address the problem

of how to retain the pleasure of seeing cows and calves, pigs and poultry, ewes and lambs as a pretty background to their country walks when the very survival of all these animals depends on the carnivorous element in society. Were we all suddenly to become vegetarian, what on earth would happen to all the farm animals — how would we maintain them and dispose of them? Think of the cost of treating thousands of geriatric animals long past their prime! And what do we do when they've all gone with only the horse to break the monotony of the vegetarian landscape, most of which will have returned to scrub?

The more adventurous members of the Vegetarian Society have come up with answers to these posers. Why not turn our farms into a patchwork quilt of national parks with the farmers as park-keepers? Is this bad news for the budding organic brethren who need all the extra FYM they can find to grow their crops? Well no, not if we harness the organic potential of household and factory. The cost of all this and whether there'll be enough of it, i.e. a feasibility study, has yet to be carried out.

There is then the small problem of how to control the breeding of the various livestock, what ratio of young to old to maintain and what to do with the geriatrics that are no longer a pleasant sight. Wildlife parks, apart from animals which are bred for sale or exchange to other parks, often use hormone treatment to prevent breeding. Such treatment may not be acceptable to Green vegetarians who visualise that all animals should express their natural propensities. When the time comes for the geriatrics to 'move on' are they to be buried, cremated or converted to fertiliser for organic croppers? The veterinary profession may have to change gear to overdrive, or re-educate themselves in homeopathy. The fertiliser and agro-chemical companies would either fold up or change direction towards organics and bio-technology.

This revolutionary change is considered to be a reasonable and rational alternative without any thought being given to whether the public at large wish to change their diet in this way, or whether farmers would be willing to be taken over and turned into park-keepers.

The Vegetarian Society is more concerned with the

efficient use of protein. It is, in its view, wasteful to grow our own and import vegetable protein to feed to animals to feed to humans, i.e. cut out the middlemen — the animals — and the exercise becomes much more viable. Perhaps they should first turn their attention to the motor car. They would find that the internal combustion engine is not very efficient in converting expensive fossil fuel into the energy required to develop forward motion. Walking from Land's End to John O'Groats is a much more economic method of propulsion than going by car — but there are other considerations, and it is these that the Vegetarian Society is quite happy to totally ignore.

To return to this question of anthropomorphism — the investment of animals with human feelings — we are all guilty (if guilty is the right term) of this kind of behaviour. We talk to our pets and even answer for them. People living on their own find great comfort in their pets which never utter a word. The pet becomes a replacement for a human being, and is often a much better choice, but rather than attempt to behave like the pet, the owner 'invests' the pet with human attributes, behaviour and responses.

This behavioural syndrome is well illustrated by some members of animal welfare groups who, with only superficial knowledge of life on the farm, presume to dictate to the farmer how he should treat his animals on the principle that what is recommended conforms to the way the persons themselves would choose to be treated. In some cases there is justification for complaint. Every system has imperfections and if there are faults in the process of production, the answer is not to shut down production but to identify the faults and seek to eliminate or minimise them. Incidents of cruelty on the farm are due to incompetence, insensibility but probably most of all to greed.

Throughout agricultural history there has always been a strong political lobby to produce cheaper and cheaper food — not just to grow two blades of grass where only one grew before but to keep two or more animals where there is only room for one, to treat animals more like robots. It is not a big step from a moderate increase in stocking rate on a pasture to the type of intensive pig and poultry keeping which today is in dispute. Vegetarians have joined with welfare groups in objecting to all forms of intensive livestock production.

Intensive livestock production did not materialise by choice, but by necessity — the need to provide cheap food and earn a reasonable return. Whether we like it or not, livestock are commodities as in any other business, and ways and means will become or be made available to increase profits by manipulating these commodities. Any system which results in the deterioration of the commodity is unlikely to be profitable, so it behoves the farmers to ensure that their livestock are healthy. The farmer is in business to make money not to satisfy the demands of well-meaning members of welfare and vegetarian societies whose ideas of slowing down production would coincide with lower farm incomes.

The difficulty comes when having to choose how much freedom an animal should have. Pigs and poultry, given the choice, will move indoors in bad weather (as will cattle and sheep if the facility is there). Outdoors, apart from inclement weather, there are problems associated with disease, parasites, predators, bullying, additional feed and inconvenience for both animal and operator. It seemed sensible to bring pigs and poultry indoors, the main disadvantage being the capital cost. The initial decision of how much space to allot per animal had to be arbitrary, and on the whole the procedure has been a success. It is where congestion has occurred that welfare groups have been justified in objecting, e.g. veal crates where calves cannot express their natural movements, cages with two or more birds encouraging bullying (it can only be minimised, not eliminated) and tethering of sows producing arthritic limbs. Such faults are totally unnecessary and can be put to rights.

Animal protection groups unfortunately illustrate their opposition by showing selected pictures of animals which are either crippled or in a very poor state of health. They do not indicate what the percentage is of such animals in any grouping, but clearly if any profits are to be made the percentage cannot be very high. The implication is that such instances cannot be seen outdoors. This is far from the truth as a more comprehensive survey would indicate. If a visitor from another planet arrived here to observe humanity and noted the number of crippled and unhealthy looking individuals living in a so-called civilised society, would he be justified in reporting back to base that Earth-people were cruel to their members?

With better health control and depending on soil type and local weather, pigs and poultry are moving outdoors again. Many of the intensive livestock buildings are looking the worse for wear and are no longer worth maintaining. Disposal of waste is also becoming a problem, particularly now that pollution risks the imposition of fines and nitrate levels in watercourses are being more carefully monitored. Where land is in short supply, as in Holland, the problem of dung and slurry disposal has reached a peak. The Green lobby in Switzerland has been strong enough to ban all caging of birds and it is likely that from 1992 onwards increasing numbers of animals will be 'free ranging' as welfare lobbies become stronger. This move must be acceptable providing the animals remain healthy and a reasonable income is maintained. What cannot be acceptable is the rapid development of any system which would terminate the farming of animals. This is not to say that future generations will never become vegetarian or largely so, but the process should be very slow so that whatever system is arrived at, it is acceptable to the majority — i.e. it becomes the norm. In such circumstances the change in the landscape, particularly the animal part of it, is anyone's guess.

In developed countries there is a wide range of choice for the consumer which is not the case in the Third World where protein is likely to be in short supply particularly among the impoverished. Any general reversal to a vegetarian diet in the West is not likely to be of any value to people less advantaged. Agriculture involves an inter-relationship between humans, farm animals and crops organised in such a way as to sustain humans nutritionally. When the system breaks down due to climate, geography, economics or politics, shortages of livestock or crops may occur resulting in famine and human suffering. Any substantial move towards vegetarianism could upset what is essentially a biological balance, particularly if the only source of fertiliser for the crops is derived from farm animals.

SPECIALISM

> *'All animals are equal, but some animals are more equal than others.'*
>
> (George Orwell)

Throughout the whole range of different societies in the world, certain regulations have been adopted which prescribe what food should or should not be eaten. This I have termed 'Specialism'.

We never really got into the habit of eating each other (if historians have been honest) like certain African and other tribes did in order to capture the power, cunning and virility of their enemies. This 'power' transfer or lack of it was also evident in a reluctance to eat the hare because of its timidity.

Whilst environment, we have seen, changes only slowly over the years, there are parts of it which remain unaltered not because of immobility but because of human prejudice and persuasion. Restrictions on certain foods were enforced to prevent shortages, usually animal proteins that were not readily available. Other restrictions were introduced to dictate what should not be eaten, the taboos, which are still retained at all levels of society and are often associated with religion. The origin of most taboos is somewhat obscure but is unlikely to have been arbitrary. Some experience of having eaten certain foods before a decision to ban them from the diet could well have been a precursor but reasons for such decisions can only be surmised.

Taboos on certain foods are a distinguishing feature of particular groups in society, setting them apart from others. This distinction was probably initiated to establish a kind of Specialism, a desire to indicate the relative 'purity' of the adherents, the foods proscribed being regarded as 'impure'. In primitive societies with limited knowledge, superstition was part of the life-style and a close relationship was maintained with a god or gods who were assumed to supply the bounty of food. Thanksgivings were offered, usually an animal which was either burned with fire or consumed by the high priest of the sect. In contrast, the North American Indian, rather than giving thanks, prayed for forgiveness after taking a buffalo to supply his needs. Many people still say grace before a meal and the Eucharist is offered as a form of thanksgiving.

The most familiar taboos which are still common today are concerned with adherence to the Mosaic Law, reported to have been instituted after the Flood, and prior to this event the human diet was largely vegetarian. Distinctions were made between animals in relation to their 'naturalness' in the environment. Sheep and cattle were considered to be natural grazers of herbage because they chewed the cud and were incidentally cloven hoofed. These animals were therefore fit to eat. The horse, pig, rabbit and camel were also grazers but either did not chew the cud or 'divide' the hoof and were considered 'unnatural' and therefore 'unclean'. Similarly fish without fins (shellfish), land animals without legs (snakes) and carnivorous birds were all considered unfit to eat. Poultry, most of which are omnivorous, escaped this distinction.

The animal which provides an important source of protein throughout the world and not so very long ago was to be found in single pens on nearly every UK farm and close by every labourer's cottage, is the pig. It has been domesticated from the time soon after Homo Sapiens appeared and it is this animal that Jews, Muslims and some other sects avoid like the plague.

There must be some rational reason for this apart from its lack of cudding and hoof structure. The idea of a God of Jews or Gentiles creating animals and then deciding quite arbitrarily to put a curse on a particular animal does not illustrate the omniscience of the Godhead. The pig, left to its own resources as it often was in primitive societies (and still is), is a rather grubby, smelly animal using its snout to uproot anything, eating anything and everything it takes a fancy to and at that time, in the Palestine sun, had the habit of covering itself in sand and mud to protect its thin skin. It scavenged in close association with humanity and was not a pretty sight (nor for that matter were some of the humans). Perhaps more important is the fact that veterinary knowledge was limited or absent and such things as worm treatments were unknown. There are some diseases which are transmissible between humans and animals called zoonoses. As far as man and the pig are concerned certain species of both tapeworm and roundworm are common to both because of the fondness pigs have for human waste.

Either man or pig must originally have started the ball rolling but it's really a hen/egg situation and it is the human who is likely to suffer most from this relationship. Cysts of the flatworms Taenia and Cysticercus find their way into various parts of the pig, and if pork is improperly cooked, i.e. is 'measly', the parasite is transmitted into the human body and particularly in the case of Cysticercus the resulting movements of the larvae through the human body can at best cause considerable discomfort and at worst result in fatal brain damage. A similar relationship occurs with the roundworm Trichinella causing muscular disorders in man. Such zoonoses still occur in some countries which lack basic hygiene. There may well have been some experience of such infection and the more civilised members of society who specialised in religious pursuits no doubt decided that the pig was 'impure' and banned it from the diet.

The adherence to this distinctive denial separated the Specialism of the 'pure' or righteous from the commonness of the 'impure' or abominable and the fact that this stricture has survived for thousands of years is an indication of the powerful influence that a conditioning programme can have which has been consistently instilled from the cradle. There is no element of animal protection in this denial. The pig was, and still is, regarded as something rather nasty by these groups. A New Testament report emphasises this fact when devils, after being cast out of a madman, were deliberately transferred by Jesus into those Gardarene swine resulting in their proverbial rush to the sea. This report appears to conflict with the recommendation in Mark 7:15 that 'There's nothing outside a man which by going into him can defile him' — a timely signal adjudged by the Gentiles (and other abominables) to cancel out all those inconvenient food restrictions in Leviticus.

In India there are well over 100 million cows, animals

which are uniquely protected as a kind of sacred trust for the very simple reason that they provide the most wholesome and inexpensive food known to man — milk, involving no slaughter. It is understandable that in a country where there is abject poverty, such animals are held in high esteem, a respect which compares with the way some organic farmers feel about the soil. In that country too, there is a kind of Specialism which divides society. Whilst the whole population shares the same environment there are elements in it which pose a risk to certain sections. These elements are called the Untouchables — human beings who have only to look at a plate of food about to be consumed by a Brahmin to make that food uneatable. Any attempt to improve the environment in such a scenario is submerged by the pressures of the caste system, part of the country's life-blood which caricatures the idea that some people are more equal than others.

The unlimited propagation of vegetarianism, it is suggested, is not good for the environment. To this can be added the belief that the propagation of Specialism is likely to lead to imbalance. There is no virtue in any society which expounds the idea that the environment should be divided in such a way that certain sections automatically claim the best and there is then an apartheid-like gradation down to a level of society which is regarded as an environmental embarrassment.

6

THE CARBON CONTROVERSY

'All things are artificial,
for Nature is the art of God.'
(Sir Thomas Browne)

In 1881 Charles Darwin published *The Formation of Vegetable Mould through the Action of Worms*, a reprint of which was made in 1945. During the 1930s and 40s there was a surge of interest in farming operations, particularising the role played by the soil. We had *Famine in England* and *Alternative to Death* by Lord Portsmouth, *Look to the Land* by Lord Northbourne, *The Rape of the Earth* by Jacks and Whyte and *The Living Soil* by Lady Eve Balfour — all issuing warnings about playing fast and loose with the soil. In the thirties Sir Robert McCarrison had pointed the finger at nutrition as being the essential basis for national health related to a specific type of agriculture. The outcome of all this interest was the foundation of *The Soil Association* initiated by Sir Albert Howard. The Body Organic had been born but it was to be nearly fifty years before the roots of this association began to proliferate and to become instrumental in completely changing the UK policy from one of maximum production to that elusive quantity optimum production.

The preceding five chapters have merely been an introduction to what follows — a subject which has become a regular conversation piece in farming circles — whether to go organic. There has always been controversy as to what in life is genuine compared with what is merely a copy, and it is this type of argument which has infiltrated into the discussion of what people regard as genuine in farming. It all focuses on the question of what we mean by nature in the environment.

God's nature is rarely what people imagine it to be. It would certainly appeal to the biologist obsessed as he is with wildness, but most people prefer the artificial environment man has created but which is not often recognised as such. Wherever we look today we see the influence of man on the environment and the general impression on the whole, certainly in this country, has been that this artificial environment is 'good' i.e. pleasing to the senses. Enough poetry has been written about the countryside to fill any moderately sized library. 'Come friendly bombs and fall on Slough . . .' is probably a sentiment of John Betjeman's shared by the majority, but Slough is hardly the kind of countryside we have in mind.

Most people then were relatively content with the environment until a government department was formed in the 70s to protect it from the ravages of mankind. In no time at all we suddenly began to realise how awful the countryside was becoming. We hadn't actually noticed this change ourselves but so many knowledgeable people began writing about it, the newspapers were full of it, and television programmes told of the disastrous consequences that would ensue if all this agricultural rape was allowed to continue.

Old words and expressions began to take on entirely new emotive meanings — natural, wholesome, organic, pure, farm-fresh, unadulterated, health-giving, low-fat, Green and environmentally friendly — all trotted out regularly, the main function of which was to provide salespersons with the kind of ammunition which would help them more readily to sell their products. We have now had a surfeit of these descriptions, so much so that we assume virtually everything we purchase possesses these characteristics.

The one word among those listed which has become really significant in the nineties and looks to be a certain twenty-first-century winner is organic. In the sixties it was a word which provided almost orgasmic pleasure to drop-outs and escapees from Sodom and Gomorrahs seeking the Good Life. It has a completeness and finality about it which conjures up visions of Fecund Mother Earth, Goodness, and Wholesome Health. What it is actually defined as in the dictionary is relating to a substance containing carbon, but we have to thank those dreaded multi-national fertiliser and agro-chemical companies for bringing this word into its well-earned majority.

A little more chemistry now for those who can't recall or have never had the chance to recall.

Carbon, so essential to the meaning of organic, is a non-metallic element which, whilst being in limited supply, forms more compounds than all the other elements put together. Scientists at the early junction of knowledge thought that carbon compounds could only be obtained from anything living or that had died and so called them organic. There are so many of these compounds that a separate department was formed to study them — organic chemistry as opposed to inorganic, i.e. without carbon. The distinction between the two wears a little thin when organic carbon atoms combine with the inorganic metals iron and magnesium atoms to form respectively haemoglobin and chlorophyll — but for most purposes organic can be regarded essentially as containing carbon. People using the word organic imagine everything about it, including the constituents, is sweetness and light, until closer examination shows that this is not so. Carbon monoxide, that highly toxic gas emitted by incomplete combustion of carbon dioxide (CO_2) in the cylinders of car engines is anything but beneficial. And then, of course, there is CO_2 itself, essential to trap the radiation from the earth to maintain a warm atmosphere, but too much of it and we move nearer to the temperature of Venus which has an atmosphere of almost pure CO_2 and a temperature of 500°C to go with it. The greenhouse effect is considered to be the result of a combination of human activities which increase the levels of CO_2 preventing excess heat being radiated from the earth's atmosphere thereby increasing global temperatures. Other gases like the chloro-fluoro-carbons do little to help maintain the protective ozone layer. Decisions are currently being made to try and offset these effects, but there are some suggestions of what we should do which would involve enormous cost. The adage that 'it may never happen' is not one supported by some scientists, but economists operate on relatively short-term programmes and to try and cater for our great-grandchildren and beyond may be an insurance that is financially unacceptable. There are already indications which suggest that the heating-up process might have been overstated and that for a given temperature rise, the time involved might be hundreds of years rather than decades.

But to return to organics — what is this controversy that has arisen? Diamonds, we know, are virtually pure carbon competing in value with those inorganic metals gold and platinum which have the same mercenary quality but not the romance of diamonds. Mention plastics, however, and the hair rises at the back of the neck — the materials composed of long strings of carbon polymers which compete with diamonds, at the other extreme, for notoriety. Plastics have exacerbated the throw-away habit of humanity and the fact that they are nearly indestructible is plain for all to see.

No, the real controversy that we are concerned with relates to the production of food, the increasing cost of this process using man-made chemicals to stimulate or protect crops and livestock, and the dangers inherent to consumers claimed by certain members of pressure groups. The term 'organic' is now used to describe food which has seen neither artificial fertiliser, nor herbicides/pesticides of any kind on its way to the dining table. To the Greens such food is environmentally friendly, being produced with a minimal use of scarce resources such as fossil fuels needed to produce fertilisers and agro-chemicals. Inorganic food, which comprises about 90% of our diet, requires this conventional treatment in production. The Greens have gone one step further, claiming that organic food is superior to inorganic, originating from the goodness of the earth and lacking any residues of man-made chemicals. Now admittedly some of these claims have gone a little over the top, but the fact is that scientists trained to distinguish truth from fiction have banded together to instigate the formation of nationally and internationally recognised organic groups and consumers are falling over themselves to buy the produce from these groups. This phenomenon is then not just confined to the UK, but is world-wide.

It was in 1563 that Bernard Palissy remarked: 'You will admit that when you bring dung into the field, it is to return to the soil something that has been taken away.' The value of animal waste in promoting plant growth was already evident. In 1912, Sir E.J. Russell noted that whilst plants grow satisfactorily with inorganic nutrients alone, in natural conditions 'their nutrition always proceeds in the presence of organic matter.'

It was really the period from 1930 onwards, when an increasing interest was being taken in the use of artificial fertilisers, that a new concept was born. It was the concept of holistic farming introduced by Sir Albert Howard and later supported by Lady Eve Balfour, and which is now resurfacing. The soil is considered to be something more than a medium in which to grow plants. It is, rather, a complex world of biological minutiae, of dead and living organisms which provide a natural balance of nutrients — the humus — by which the plants grow and flourish. These nutrients are basically organic and the logical way to stimulate more plant growth is to augment the supply of organic material — the waste from town and country converted into compost — a material which can be readily incorporated in the soil to provide the necessary humus and possessing both the physical and chemical properties needed by crop plants. The use of artificial fertiliser is considered to interfere with this natural process and in addition, the use of agro-chemicals (pesticides etc.) is regarded as even more deleterious to the association between the soil, the crop and the environment. People who originally became interested in these ideas grouped together to form the Soil Association which became the official voice of organic farming.

It attracted relatively limited interest at the time because the UK was soon to become involved in the Second World War when inorganic fertilisers came into their own to feed a nation under blockade, an achievement which could never have been attained by the sole use of organic fertilisers. After the war, the fertiliser and agro-chemical companies took on a new lease of life — until, that is, they became as it were 'hoist with their own petard'. 'Me-thinks they have produced too much' might have been said about their efforts. We have now had embarrassing food surpluses for a number of years and it was under these circumstances that attention was drawn towards farming systems which did not demand high chemical inputs, which were not concerned with maximising production and which released farmers from what has been described as an agricultural treadmill. The Soil Association was a body already experienced in such systems to which serious attention was to be paid during the 1980s.

A scientist who would be more properly regarded as belonging to the inorganic school did begin to question the use of fertiliser in the sixties. André Voisin declared: 'The use of mineral fertilisers is one of the greatest discoveries of modern times.' But at the same time he appreciated the imbalance that could be brought about by its improper use, particularly in relation to trace elements. He realised that the use of fertiliser had become so commercialised that the primary aim was yield rather than the preservation of the soil and the biological quality of the produce.

This obsession which organic groups have concerning the soil as a living biological entity makes one believe that soil is absolutely essential to grow any plant. This however is not the case. In early studies of the nutritional requirements of plants it was found that growing them indoors in solutions containing varying amounts of nutrients was an ideal way to be able to have virtually complete control over the plant environment without interference from extraneous sources. The main problem was to maintain the plant in its natural upright position and this was achieved by using gravel as the basic stratum to hold the plant *in situ*. Solutions of fertiliser containing the main plant nutrients, nitrogen, phosphate and potash (NPK) together with trace elements like sulphur, magnesium and calcium could then be added, and such solutions can be used indefinitely, providing the required levels of the constituents are maintained.

This system, called hydroponics, has been commercialised and is appropriate where land is in short supply as in

Holland. The capital outlay is high, but subsequently the system compares favourably with any other means of production. In the region of 80% of total horticultural produce is grown by this means in Holland, most of which is exported. It represents the end of the line as far as inorganic farming is concerned and the produce appears to be identical with that produced either conventionally or organically. Gravel is, or certainly can be made, a sterile medium, so the input of agro-chemicals to deal with soil-borne pests can be modified. This type of specialised production is a far cry from that discussed in what has been called *The Dutch Report on Alternative Agriculture*. The authors here are concerned for the future well-being of agriculture and search for an answer by looking backwards, recalling the historical reverence for the cosmos uncontaminated by the interference of modern technology. The assumptions made in promoting both conventional and alternative agriculture indicate a requirement for much more research.

Hydroponics is understandably a very specialised system of production and is unlikely to have much impact on the organic/inorganic controversy. It does, however, illustrate fairly conclusively that soil is not necessarily the life-blood of agricultural production.

In 1975, at Wye College, a conference assembly concerned with biological husbandry heard the following statement: 'The complex system represented by plant roots, associated microflora and soil organisms . . . which has developed into its present complexity over millions of years of evolution cannot be short-circuited by chemical means without risk or damage to the health of the soil and of the plant.' Now you would surely think that to support such a forthright challenging claim, the speaker would have indisputable evidence which would once and for all ring the death knell of chemicals. Not so. Indeed rather than taking millions of years to develop an acceptable organic system, recognised organic groups reckon they can do it in about three years. Such intellectual exercises are all very well if they provide meaningful answers, but to promote organic farming by trotting out statements which can in no way be substantiated is unlikely to have much impact on the more perceptive inorganic farmers.

In biological terms, all agricultural systems interfere with the natural processes so cherished by organic groups. Is there something inherently valuable in these natural processes? A field of weeds is the result of a natural process but has no value apart from a facility for biological study. The development and increasing use of fertiliser and agro-chemicals exchanged biological systems for non-biological systems whilst the move towards organic farming is this exchange in reverse. This claim by the organic lobby is only partially true. Crops grown by conventional (inorganic) means are dependent on the biological system of nutrient uptake whatever the source of the nutrient. Similarly organically grown crops depend on non-biological inputs, cultivations etc. for successful growth. This kind of exchange is not so remarkable. Humans have exchanged the biological activity of walking and running for motorised transport to facilitate better communication, using the horse and cart in the process. This achievement involves a fairly profligate use of fossil fuels, but few organic growers or members of the Green Party would be prepared to dispense with their cars and revert to the horse and cart.

A kind of organic system developed by Rudolf Steiner in the fifties was called bio-dynamic farming. Steiner considered that the solar system, plants, animals and humans have a kind of anthropomorphic relationship by which the activity in one dimension could influence the activity in another. Thus times of sowing and harvest should be related to lunar activity. The system was something more than exchanging one kind of manure for another and was a move against the concept of producing more and more. Produce from bio-dynamic farms was named as such and commanded high prices in the market. Did this illustrate the gullibility of the misinformed consumer or was there really something of value in this system? A German farm, managed bio-dynamically since the 1930s, using neither artificial fertiliser nor pesticides and limiting bought-in feeding stuffs, still operates successfully, the output comparing favourably with that of conventional farms. The essence of the success is based on many years' experience on a particular farm in a particular climatic area, and the information from such a farm can make a valuable contribution where similar developments are contemplated in a similar environment, but such results are not readily replicated. Ideas like these have not, however, been very popular in practice, and currently there is only a small number of organic farmers in Germany, but interest is growing.

It is certainly growing in the UK and D. Strickland of Organic Farmers and Growers is currently looking for 100 more organic farmers to meet the demand for organic produce. It was early in 1989 that organic farming was officially launched and regularised by the publication of the United Kingdom Register of Organic Food Standards (UKROFS). Within its seventy-four pages are very precise instructions for both farmer and processor of organic produce, and those who adhere to the rules are qualified to attach the Soil Association symbol of purity to their produce.

In this booklet, organic farming is defined as a system 'which works with natural systems rather than seeking to dominate them and which will produce optimum quantities of food of high nutritional quality by using management practices which aim to avoid the use of agro-chemical inputs and which minimise damage to the environment and wildlife.' In addition the use of fertiliser is banned. It is a truism that agricultural practice of any kind is constantly in competition with certain elements in the natural environment. It is the degree of intensity and direction of such practice which locally may have a more dramatic effect on the ecology of a particular area. Agriculture in the UK has evolved from an environment which originally was predominantly wooded (which suggests that we may be ideally suited to redevelop woodland as an alternative form of biomass). The disappearance of this woodland will not have had the same effect on CO_2 exchange which the disappearance of the world's tropical forests is having, nevertheless, there have been considerable changes in our own environment and wildlife due to agricultural intrusion. Whether such intrusion can be described as damage depends on what the perspective might be.

The notion of producing optimum quantities is not defined. Presumably the optimum would depend on market demands. We cannot however assume that the current farm surpluses will last for ever. As populations increase and developing countries raise their standards of living, there may well be an increasing demand for food. Already among organic groups there is this unresolved problem of whether to claim that organic farming can equate to that of conventional farming, or to advocate organic farming as a means to control surpluses. Certainly if there were to be increased demands for food, organic farms would be hard put to it to meet them. The whole idea of organic farming is to take the pressure off, to remove the treadmill and to establish a more relaxed system of farming. Unfortunately a farming system cannot be evaluated in isolation. There are outside pressures on the farmer — he has to make a living and like anyone else would be reluctant to lower his living standards in a competitive world. Currently, the competitiveness of organic farming depends on the premium the consumer is prepared to pay for organic produce. If conditions change — if reliance on the energy from fossil fuels is no longer vital, if the consumer becomes sufficiently informed to realise there is no more intrinsic value in organic produce, if the imminent fear of global warming recedes — then these premiums might become a thing of the past and organic farmers would have to compete in real terms. The situation at present is that there is a two-tier market, the one for the majority of consumers which is inorganic, and the organic one for the minority. This does not detract from the value of organic farming, but does characterise it as specialist production.

The directive in the Register concerning veterinary treatment allows conventional drugs to be used only 'when it is necessary to save the life of the animal . . . to prevent suffering or . . . where no alternative treatment is available.' In general the veterinary profession would agree that prevention is better than cure, but they would go further and say that preventive medicine is better than curative. This is not acceptable to UKROFS unless the drugs are organic. It is possible to visualise a situation where the animal reaches the stage of requiring conventional curative treatment because conventional preventive treatment has been withheld. Is it reasonable to expect the farmer to have to choose between a regime which promotes animal welfare and one which adheres to a list of regulations?

It is one thing to produce a set of rules so that organic farmers and processors together with consumers are precisely clear as to what is being offered in the marketplace, but it is quite another to monitor this production. Inevitably a certain amount of trust is needed to make the system operative. Monitoring the production of imported organic food is likely to be much more difficult. The consumer has no irrefutable guarantee that he or she is getting what is being paid for — exclusively organic produce — and should, therefore, realise that an organic farming system is not infallible.

The principles expressed in UKROFS might be regarded by some as being more idealistic than realistic, originating from concepts which whilst being something more than armchair thinking are likely to be difficult to practise in the field. Arable farmers are always going to be at a disadvantage in having no ready source of organic matter. Reliance on this material from outside sources may be necessary and these sources can only be on inorganic farms — organic farms will retain their own sources.

The organic precept not to dominate the environment is rather overshadowed by the environmental element of weather which dominates most farming activities. If in fact the environment was 'left to itself', it would quickly establish its own dominance. Working with nature is not easy when many of its biological constituents appear to be set against agricultural practice. There are times when the farmer has to dominate as a means of survival. There are other times when 'co-operation with nature' is called for. It is the admixture of these apparently opposing commitments that has produced the kind of varied landscape we see around us today.

What is sometimes forgotten in advocating organic farming for general practice is that we live in a temperate climate in which biological activity is comparatively subdued compared with that, say, in the tropics. The kind of protection recommended for organic systems in temperate zones would be quite inadequate in the tropics. The mild processes involved in homeopathic treatment would have little effect for example on the ravages of the locust, the mosquito or the tsetse fly, difficult enough to control by inorganic means. In temperate zones, the current position seems to be that more research is necessary to arrive at better assessment of organic farming and there is no clear indication of a more profitable enterprise except in certain localities and under a given set of circumstances. A widespread conversion to organic farming does not appear to be feasible and one is led to ask whether organic farming is confined to the type of environment which allows it to operate successfully. To achieve this generally might require population constraint since by definition organic farming is extensive rather than intensive so that any increase per unit area is likely to be limited.

Most farmers are fairly practical people having little spare time to commune with nature. Not so people outside the industry who purport to concern themselves with environmental issues. Farming has been described both as an art and a science. To the writer and the painter it is the art in farming which provides the interest. Writers like Wordsworth and Keats, painters such as Constable and Turner were not concerned with the daily exigencies which press on every farmer.

Encompassed by this kind of poetic and visual halo there are some people who regard farming with the kind of mysticism which sees it only as a wonderful element in nature. The soil, being a rich conglomeration of biological activity, has acquired a sacredness of its own and needs both respect and understanding from humans to be able to reveal its secrets. Religious beliefs have helped to promote these attitudes and the use of man-made chemicals which adulterate the soil in the process of farming is regarded as sacrilege. A relationship is developed with the soil which jealously defends it from 'outside interests' that seek to exploit it and use it as a means to an end. It is a relationship which is very satisfying, enough in fact to convince these people that the soil has an esoteric life of its own and that any kind of cultivation is a form of abuse. Where such abuse or interference is intensified this esoteric life will be ravaged. Now these attitudes, whilst appearing way out, do exist, and may account for some of the apparently irrational ideas stemming from organic groups.

The majority of farmers who do decide to farm organically do not necessarily do so for altruistic reasons. They do not, however, anticipate making more money, or growing larger. The pressure to produce more has eased because of surpluses and there is an increasing demand for organic goods which command a premium. To some extent these premiums compensate the farmer for lower production. The increased labour requirements are balanced by lower inputs (of fertiliser and agro-chemicals).

The success achieved by one organic farmer is not readily replicated. At entry, he needs to be financially sound, i.e. not at the wrong end of a large overdraft. A family farm is likely to perform better than one employing outside labour — organic farming thrives on certain attitudes which are better developed in a family situation. Livestock farms would be easier to convert than arable — where all the organic manure must be imported. The expertise with both stock and crops needs to be much higher in organic systems which do not have the same control of pests and diseases as inorganic systems.

It has been suggested that some farmers opt for an organic system wishing to escape from the treadmill of fertiliser and agro-chemical escalation — the kind of pressure which is a norm in the city rat-race. The case, I believe, has been overstated. The farming way of life is a kind of treadmill in so far as there is repetition year in, year out. Whilst the pattern may change within the system, the system is still repetitive but rarely loses its charm. The real treadmill is a financial one, trying to match increasing costs and decreasing returns, a headache for both organic and inorganic farmers.

The results which have been reported from organic farms in this country indicate that the best of these can compete with the average of good inorganic farms (but not the best). The output per hectare generally is lower which is to be expected in an extensive system. The major problem on organic arable farms is weed control. Particular attention must be paid to times of ploughing and sowing and there is usually a need for more harrowing at optimum intervals to weaken weed development. Control of pests is rather in the hands of the gods and attacks for instance of mildew and blight on vegetable crops can be overwhelming.

One interesting test case for the organic farming system is Rushall, a farm of 1,650 acres in Wiltshire. The owner, Barry Wookey, decided to go organic in 1970. After reading his book *Rushall, the Story of an Organic Farm*, I got the impression first, that the owner of 5,000 acres was in no way financially pressured when a change to organics was implemented. There is no reported consultation with the bank manager concerning the wisdom of embarking on a financially suspect enterprise, and any possible failure of the system was not going to be too critical. Secondly, bearing in mind that there was no financial pressure, it would seem that the organic farmer gets far more pleasure from his business than the inorganic farmer, because he is up against odds without chemical aid and with only ingenuity to see him through.

Some of the facts emerging from further study of this enterprise are:

1. Prior to entering the organic system any particular field was 'blasted' (my expression) with agro-chemicals to ensure a relatively clean entry.
2. The difference in treatments necessary between fields, not just as between light and heavy soils, to establish a satisfactory organic system and the extended fallowing necessary to contain weed establishment.
3. The use of older, taller grain varieties which suppress weed establishment by reducing access to light.
4. This particular organic system relies on inorganic inputs:
    (a) Calves from inorganic farms.
    (b) Foals (origin not stated).
    (c) Sheep which graze keep receiving NPK.
5. The success of the enterprise has also been helped by a by-product of the farm — organic bread, and by visitors and publicity.

The system at Rushall suggests to me that integrating

organic and inorganic systems offers more possibilities than adherence to a rigid set of rules. This is a point I shall refer to later.

Results from organic farms in other parts of the world tell a similar story. Information coming from West Germany, France, Switzerland, Austria, the Netherlands and the United States indicates a major reliance on organic material from other sources. All arable farms wishing to go organic must have a source of FYM within transportable distance. The source is exclusively from inorganic farms where the organic material is an 'embarrassment' and is usually collected at no cost to the organic farmer. With such free inputs, any assumptions made on the output of organic farms must be qualified.

Perhaps the most valuable contribution that organic farming could make is to reduce the incidence of soil erosion. The problem is not so serious in the UK and Europe, although in the early 1900s corn was described as the 'syphilis' of Europe because of the depletion in soil structure and erosion caused by the kind of tillage operation which compacted the soil on sloping areas. In 1940, Jacks and Whyte showed the extent of soil erosion in North and South America, Australia, Russia, South Africa, China and the Middle East, all resulting from denudation and excessive monocropping. Vast areas like the Dakota Dust Bowl are unlikely to be changed much by organic farming, but in smaller more confined areas where animals are available, it can help to prevent or restrict erosion.

In the USA the rate of topsoil loss is calculated to exceed the rate of regeneration on at least a third of the arable farms, and more sustainable farming systems are being sought. Since only about 3% of the population supply food for the remaining 97% any increased costs involved in changing the farming systems must be borne by the majority. Current studies have shown that changing to organic or sustainable systems incurs an overall cost increase of about 10%, most of which is borne by consumers willing to pay extra for organic produce.

How much do consumers know about organic food? In 1972 in the US a questionnaire was sent out concerning food and health, the answers to which were either true (T) or false (F).

1. Chemicals reduce the value of food (F).
2. Man-made vitamins equal natural vitamins (T).
3. Processing and refining of food reduces its nutritional value (F).
4. Chemical sprays endanger health even when carefully used (F).
5. There is no difference in food value between that grown on rich soil and that grown on poor soil (T).
6. Shipping and storage reduces food value (F).
7. Food grown inorganically is as good as food grown organically (T).

Replies from the general public gave 43% correct answers, but those from a specific health group gave only 14% correct answers. The propaganda for organic food had been clearly very successful, but the answers generally showed that consumers were not particularly well-informed, and there is evidence that outlets for organic food are increasing.

In the 1980s a similar type of questionnaire was carried out by Elm Farm Research, an organisation which took over Haughley Research Farm, Ltd, a project conceived by Alice Debenham in the 1930s.

Q. Are chemical sprays and fertilisers appearing in farm crops?
A. **Yes** 65% **No** 14%

Q. Is the environment being destroyed by today's farming methods?
A. **Yes** 60% **No** 12%

Q. What is organic farming?
A. **Know** 28% **Don't Know** 72%

Here again the organic lobby has done a first-rate job of either misinforming the public or allowing the public to be misinformed to further their cause. In a survey, some 77%

of respondents said that they would buy organic food if it were more available. What is alarming is that people with only superficial knowledge have the ability to influence important decisions made on the basis of such questionnaires.

The majority of the world's farmers are small-scale, i.e. under 10 hectares, and are already operating organic or near organic systems. In developed countries agricultural production is near the maximum so that any research to increase production might well be better directed towards these smaller farms in countries where the population increase is highest and where the additional food will be needed. It has been estimated that during the twenty-first century agricultural production will need to double every eighteen years to keep pace with population increase. There must, however, at some point on the line be a limit to both of these estimates. There is no finance in these countries to pay for expensive Western-style inorganic systems and any increase is more likely to come via the organic path.

In various collected papers on organic and integrated farming systems, the authors express the need for a sustainable agriculture without indicating any period for sustainability. Is there any value in looking towards a situation so far ahead that it is barely conceivable? In the fifties and sixties there was a demand for maximum production whereas the reverse is now the case because of surpluses. Deceleration can hardly be regarded as having any permanence and whilst there is an immediate requirement to cut back, increasing populations may change the direction of food production again.

Sustainability is not a measurable parameter. It involves an element of restriction which is only assumed to have a degree of permanence. It has been conceived under pressure by environmentalists who, like any other pressure groups, need to include a degree of exaggeration in their arguments to attract attention. The result is to advocate a policy which is diametrically opposite to all business concepts, because the advice is not to cut back temporarily but to embark on an unspecific period of sustainability which will result in limiting farmers' incomes indefintely. It has only been by ingenuity and consumer gullibility that farmers adopting organic systems have been able to maintain financial security.

The value of organic farming may not be readily measurable using co-ordinates common for conventional farming. New parameters may well be needed which assess the long-term environmental effects in varying sets of political and economic circumstances. The environment has been changing ever since humanity began to cultivate, and what is acceptable to one generation may not be acceptable to another in terms of practice, ethics, aesthetics or economics, (cf the increase in vegetarianism).

*Nicolas Lampkin of Elm Farm Research Centre has defined the principles of organic farms which match those of UKROFS.

1. To draw upon local resources — home-grown feeds and legumes for N-fixation.
2. Long-term soil fertility — deep rooting rotation crops, (similar to the Clifton Park System of the 1940s).
3. Utilise organic manures to reduce N loss.
4. Produce quality food in sufficient quantity.
5. Minimise use of energy from fossil fuels.
6. Allow livestock to express their normal behaviour patterns.
7. Allow producers to develop their potential in relation to their labour.

These are all worthwhile objectives but rely for their fulfilment and acceptance on the following precepts:

a. The progressive scarcity of fuel from fossil sources. The development of fuel from alternative sustainable sources could reduce or eliminate the problem.

b. The degree of pollution by agro-chemicals and the extent of its effect on national health. There is a dearth of research of this kind.

c. The ability to produce enough. Organic growers claim superiority in quality on slender

* Now at centre for Alternative Farming and Agroecology, Aberystwyth.

evidence, but because they are constrained by restrictions, they may not be able to produce enough should the occasion arise.

The genuineness and authenticity of the organic argument has not been helped by statements made by some of the members:

1. Pesticides can produce off-flavour.
2. Pesticides can produce changes in the biological activity of plants.
3. Food residues or accidental contact with pesticides etc. contribute to genetic damage, birth defects or cancer.
4. Fertilisers can poison streams and wells leading to loss of the protective ozone layer.
5. Pesticides may contribute to a decrease in ocean life.
6. Artificial fertilisers have been linked to sterility in the consuming animal.

Numbers one to four have either been dealt with or cannot be supported by real evidence. Number 5 referes to eutrophication or enrichment of waterways by nutrients with a high phosphate content. It can lead to deoxygenation and has been a problem in the Baltic area and some freshwater areas. The process is not entirely due to leaking of excess fertiliser.

Number six presumably refers to an Austro-German experiment which showed that rabbits fed on intensively fertilised fodder crops and concentrates with a high potash content produced disturbance in the sex organs. The study was not large enough to provide any valid statistical data. To leap from suggestions indicated in such experiments to the claim made in number six amounts to an uninformed stretch of the imagination.

Organic Farming, a recent publication by Nicolas Lampkin, is a very comprehensive collection of data relating to this system and avoids the more outlandish claims by some adherents. The following points occurred to me in studying the book:

1. The author reiterates the concept that 'organically produced food can only be defined in terms of the production system.' The original proponents of organic farming in the 1930s would, I believe, have gone much further than this rather guarded definition, and it is the more positive belief held by consumers of organic produce — that it is better for them, or tastes better or is more environmentally friendly — which persuades them to buy it. There are so many different factors involved in producing either organic or inorganic food, that any sensible or rational assessment of any differences is likely to pose difficulties. Presentation and publicity are important persuaders in consumer choice.

2. Nevertheless a statement is made that: 'Sufficient evidence exists . . . to support the view that organic farming can contribute to environmental protection objectives, and to enhancement of food quality, while at the same time helping to reduce the output of surplus commodities.'

The 'environmental protection' is in line with an earlier concept in the book that organic systems prefer to use 'materials and substances least environmentally disruptive'. This last qualification would be difficult to be certain about since so many activities are environmetally disruptive. It is the degree that is significant. Ultimately any decision to use any particular material (assuming that a consideration is made) will depend on the balance between this disruption and the benefit gained.

Substances or materials produced via the use of fossil fuels have come in for particular criticism. If the energy, saved by not using such materials, could be diverted into useful or necessary projects, the denial might be worthwhile. However, no such control of alternative usage is available, so any real value of saving energy may not be critical.

The 'enhancement of food quality' suggests, without actually stating it, that the improvement will be on inorganic food. The evidence is still inadequate to support this.

The reduction of 'surplus commodities' can be achieved with little effort, i.e. by not farming at all. Mr Lampkin suggests that a 10% conversion to organics producing a 3% drop in cereal production could, by extrapolation to 100%

conversion, reduce cereal production by 30%, equivalent to the surplus. To promote such a conversion can only be imaginary. Any production from cereals-only farms (or any all-arable farm) would be completely eliminated in the absence of organic matter.

3. The suggestion that all farms are involved in 'potentially harmful practices' does not help in the analysis of production by either method. Taking the car on the road is a potentially harmful practice, but one has to weigh the benefits against the disadvantages. The car, however useful, is far more lethal than pesticide residues in farm products, but reading about fatal accidents in our 'dailies' does not prevent people from driving in the same way that reading about salmonellosis in poultry and BSE in cattle puts people off eating eggs and beef. Is the human body more, or less, resilient to chemical invasion on the inside, than to physical assault on the outside?

4. Groups of fungi, collectively termed mycorrhiza, have a special relationship with root hairs of plants allowing better infiltration and transfer of plant nutrients, and also appear to provide a positive effect on promoting a healthy plant. To what extent do inorganic fertilisers inhibit this relationship and is this significant?

It would be enlightening too to have complete analyses of plant and root systems, at various stages, from comparable organic and inorganic produce to assess what, if any, are the differences. Assessments using questionnaires on preference and taste are far too subjective.

5. A greater reliance on home-grown produce in organic systems could have some effect on the agricultural import/export ratio if organic systems increased to any extent. Whilst there might be a saving at home for the organic farmer, there may be a detrimental impact on Third World (or poorer) countries which rely on agricultural exports for hard currency. The total environmental benefit in such circumstances may then be more apparent than real.

The idea has been fostered that organic and inorganic farming systems are fundamentally different, the one concerned with a holistic view including ethical and spiritual elements, the other primarily linear. Rather than emphasise this difference, would it not be more profitable to select the best from both to establish a kind of integrated agriculture? Both systems depend on either organic or inorganic or a mixture of these inputs, both systems use the same source of energy to run their businesses, and both provide food needed by the consumer. The organic lobby cannot deny that a combination of artificial fertiliser and FYM promotes more humus in the soil than FYM alone. The majority of farms with livestock are already partly organic, but any increase in organic input might require more livestock intensity which is not desirable. Nitrogen which is a common denominator affecting all output could be partially derived from legumes. Eliminating the input of N from artifical sources altogether would leave an unfillable gap.

The accusation against fertiliser manufacture is the amount of fossil fuel required to produce it. Some recent work by Pimental and Pimental suggests that only about 5% of total fossil fuel consumption is due to agricultural activities, some half of which is used for fertiliser production. This compares with 85% used by manufacturing industry and both public and private transport. The tourist industry relies heavily on fossil fuel consumption, but there is no suggestion of curtailing this facility, whether one regards this business as more or less essential than food production.

It is clear then that any policy to save fossil fuel consumption in agriculture will have little impact on preserving a declining commodity. In this particular respect, claims by the organic lobby of being environmentally friendly by abstaining from using inorganic fertilisers, are unreal. It seems unreasonable to censure the use of fertiliser when it is so necessary to maintain the basic requirement in food production, a requirement which the organic section could not satisfy.

The maintenance of health of both livestock and crops in organic systems without recourse to agro-chemicals is a major difficulty, requiring additional skill and intellectual input. There is great interest in finding an alternate approach to that provided by agro-chemicals which are increasingly expensive and subject to consumer criticism. This new interest is in biological control applicable to the arable farmer. It is a process which relies on the dictum that '. . . little fleas have lesser fleas and so ad infinitum.' The cost of development is sure to be high but once established the maintenance costs would be far less than for instance the costs incurred by farmers growing rape — something in the order of £10 million annually.

The biological programme amounts to a big con trick which makes use of the fact that parasites of the insects to be controlled can be conditioned or 'brainwashed' to improve their ability to seek out and destroy their prey. Another ploy takes advantage of the production by insects of substances called pheronomes which act rather like alarm signals to others of their kind. Thus by spraying 'artificial' pheronomes on a crop, aphids can be disorientated and more easily controlled. Plants in the wild also produce pheronomes, when attacked by pests, which give a signal to parasites of the pests at the same time. Cultivated plants have generally lost this ability but by spraying artificial plant pheronomes crops can be similarly protected. Other plant products include allelomorphs which discourage competition, i.e. from weeds, offering the possibility of further control by means of artificial allelomorphs. Like any other animals, pests have food preferences and dislikes. Materials called 'anti-feedants' can be sprayed on to a growing crop diverting the pests to a non-crop area where they can be dealt with (probably inorganically).

An argument often used against agro-chemicals and drugs is that they are designed to attack the disease, not treat the animal. It may be a somewhat specious argument, but this kind of treatment does differ markedly from a new kind which is attracting increasing attention — homeopathy.

It was in the fifteenth and sixteenth centuries that new concepts in medicine came to be formulated. Paracelsus, a German-Swiss physician, rejected the medical beliefs of his time and began to search for what he described as 'the latent forces of nature'. He sought out 'old wives, gypsies, sorcerers, wandering tribes, old robbers and such outlaws', preferring the common Germanic language of the innkeeper to the fine Latin scholasticism of the medical authorities. It was in fact a kick against the establishment, but in his wanderings, Paracelsus possessed that unique quality to be found in the complete observer. With such ideas he made many enemies but finally triumphed with the publication of his beliefs. He was very popular with the medical students who are ever rebellious.

Paracelsus is credited with being the first physician to suggest that 'what makes a man ill also cures him' — the seed had been sown that was to take root and become homeopathy.

Homeopathy is a form of treatment which rests on the theory that like cures like, i.e. a treatment which in a healthy person would produce symptoms of the disease being treated. The theory was introduced in its present form in 1791 by Samuel Hahnemann in Germany who found that by treating a healthy person with quinine, he could produce symptoms of the various conditions that quinine is administered to cure. It was about this time that there was a continual search for a Utopian cure, a kind of 'royal touch' which would be a cure-all and easy to administer. These ideas were developed in an atmosphere of great superstition and were influenced by a witchcraft mentality. Treatments often included very large doses of whatever medicine was recommended, quite a contrast from the minute medical doses recommended by Hahnemann.

In the twentieth century homeopathy has again developed in response to the enormous market in drugs which is confusing even to the best medical practitioners. Until recently it has been regarded with great suspicion because of the infinitessimal sub-molecular amounts used in treatments. André Voisin indicated in 1964 the tiny amounts of chemicals which would affect animal health. In lambs at pasture, one part per million of copper in fertiliser can cure enzootic ataxia (a muscular condition), similarly one-tenth ppm cobalt can eradicate pernicious anaemia, where it can be demonstrated that this trace element is lacking.

There is now increasing interest by the qualified medical profession in homeopathy because of its success in curing what had come to be regarded as incurable. The accusation that most of the treatment amounts to a placebo — a cure achieved via the psychology of the patient rather than any value in the medicine — cannot be levelled in cases of animal treatment.

The veterinary profession has successfully used homeopathic treatments where conventional treatments were either unsuccessful or prohibitively costly. Calves have been treated with a non-synthetic remedy for nasal disorders, sows with a 20% incidence of stillbirths have been treated with an old Indian recipe — Caulophylum — reducing the incidence to 2%. Treatment via the water trough has significantly reduced mastitis and New Forest eye in cows. These, of course, are all isolated incidences but they do represent successful treatments of animals. They should not be regarded as being capable of replication without reference to the particular conditions on the farms concerned. Because homeopathy depends on a very subtle form of treatment, any other dominant factor which might be present could completely invalidate the treatment.

What I believe is revealing concerns the minute amounts involved in the treatment. It has already been stated that the amounts of chemical residues left behind in crops and livestock are barely measurable and unlikely to pose any risk. In view of the equally tiny amounts of chemical involved in homeopathic treatments, can we be absolutely sure that a process could not operate in reverse? This suggestion is not to join the scaremongers, but an observation that justifies further research. There have been claims that there has been an increase in degenerative diseases in most industrialised countries and attempts have been made to link the increase with pesticide residues. There is no positive evidence here, merely one of association. Clearly any contemplation of living in a sterile environment would be counter-productive, but the health of the community is paramount and the main indication is that far more government-sponsored research is necessary to study these unexplored and unexplained phenomena.

7

PROPOSITIONS

'To complain of the age we live in,
to murmur at the present possessors of power,
to lament the past
to conceive extravagant hopes for the future,
are the common dispositions of the greatest part of mankind.'

(Edmund Burke)

Any extravagant plans for the foreseeable future of farming in the UK are likely to be constrained by policies coming from Brussels. The majority of farms in the European Community are small relative to those in this country, where the larger farms are considered to be more efficient. Any decisions will no doubt be based on which considerations are rated higher — efficiency and economics or humanitarianism and tradition. There are good arguments in support of both groups and both considerations, but ultimately it will be a political decision which is more likely to favour the small farms. In the long run this may well be the right decision since it is on these farms where there is the greatest potential for improvement.

Any such proposals will not be supported by the farmers and politicians of the UK where financial support has been given to the efficient but where the rewards have been all those inconvenient food surpluses.

1. RESEARCH AND DEVELOPMENT

'Man proposes but God disposes.'
(Thomas à Kempis)

The above quotation should be brought up to date by saying that some men propose but government departments, in particular those concerned with agriculture, dispose. A decision to cut around £30 million in 1989 from the funds required for agricultural research, following the Barnes report, was not made because scientists were becoming less productive, but because government officers were becoming less efficient at balancing their books. Some smart Alick with an Arthur Daly mentality came up with the bright idea of cutting out research which had anything to do with farming. Somewhat short of vocabulary he or she called this near-market research, particularly if it showed what farmers should or should not be doing. The outcome has been that at least twenty-three research centres employing 2,000 scientists have been axed and a casual observer might assume that this is a just reward for producing too much. The scientist has been doing too well! Research which has the public good in mind and that which might benefit the farmer are worlds apart in the opinion of treasury officials. A typical example relates to the nitrate problem. The nitrates which stimulate crop growth and the nitrates which infiltrate into our water supply have no connection in the mind of the civil servant so that axing research on the one is unlikely to affect research on the other.

The rather messy outcome of the BSE and salmonella fiascos in beef and poultry have suggested to the government that a cut in research on salmonella will help to finance the escalating costs of the Common Agricultural Policy. What is really surprising and must have shocked some members of the Green Party is the decision to axe work on food safety. Clearly some discerning civil servant has realised that the human frame and contents have sufficient resilience to withstand this poisoned environment we hear so much about.

An enormous amount of research in the past has gone into studying the cost-effectiveness of agro-chemicals. Is there not a requirement now to study ways of reducing these costly inputs whilst maintaining profitability? Such a procedure would not be in response to panic alarm calls from the Green Party but in response to the need to change the emphasis from that of 'destroy' to that of 'control', from one of increasing production to one which safeguards production.

Gradually the agricultural industry will have to finance more and more research, the benefits of which will be the priority of those within the industry. These benefits may not always coincide with what the public needs and may not be the ideal way to produce the greatest happiness of the greatest number — but it does relieve the government of a responsibility and from being accused of preferential treatment.

2. SET-ASIDE

'There is no cure for this disease.'

(Hilaire Beloc)

The disease symptoms are those of excess production and excess pollution. The cause of the disease is greed, not the flagrant greed of the avaricious but the greed that is built into the ambition to grow bigger, to produce more, to compete, to make more money and to be single-minded. These are the characteristics of the average businessman trained to believe in progress but mindlessly unaware of any side-effects or consequences which might be detrimental.

When food surpluses first appeared the robotoid mind of the European bureaucrat sought an expediency. The granaries were already full to overflowing, the quick solution was to destroy all this extra food. There must be some esoteric, rational and ethical virtue in doing this which escaped the rest of us.

But a much better remedy was presently to be discovered, a treatment to relieve all that painful stress and embarrassment. Stop farming — or at least leave some of the land uncultivated. Why on earth didn't someone think of this revolutionary idea before those first supluses emerged? The problem then is solved. No more over-production, no more screwing the taxpayer to store surpluses that nobody wants, no more destruction of good food and no more selling cheap food to inept governments east of the old Berlin Wall. Well perhaps this conclusion is a little too facile. Let's take another look.

The problem was largely centred in the arable section — too much grain was being produced. We are not concerned with the average small farms of the EC but with the medium to large farms.

Government thinking went something like this: 'If we ask for volunteer farmers to take 20-100% of their land out of production it might be a better way to cut grain production than these co-responsibility levies, which are unpopular, or imposing quotas. We shall, of course, pay the farmer up to £80 per acre for agreeing to this and the land could be out of production for, say, five years, or even permanently. Something extra could be paid for turning farms into wildlife habitats.'

Now any farmer not making money in the arable sector, and there are a few of them, can sit up and take notice. Being paid by the government to do nothing and reducing those costly inputs could help to stem that terrible overdraft and ease the pressure.

The surprising outcome is that whilst a lot of interest was taken initially in the idea of set-aside, there were few who actually took up the scheme. Farmers are the kind of business people who are schooled to work by the sweat of their brows and are ever suspicious of government's armchair motives. Being offered money for doing nothing must have a catch in it. It goes very much against the grain to farm badly, which set-aside could encourage. Good farmers are not accustomed to looking over a field of weeds and then walking on oblivious to the weeds. (The few bad ones are needed by pressure groups and the media to justify their various accusations.)

There's the problem of what the neighbours might say if their crops become covered in weeds and pests, and what about the townies passing along the road and seeing the place in such a mess? The biologists and the butterfly-men might approve, but is this a satisfactory way to make a living on the land? Such a farmer might say to himself: 'I could of course go organic — they say there's money in it if you go about it the right way, but I'm not sure. The people I know in it don't really speak my language and quite a few of them have other businesses anyway. But I do have quite a bit of land that doesn't do well and some of the headlands are a bit thin — I'll have to think about it. In any case I'll have to have a word with the landlord first — I can't see him liking the idea of horses with townies on their backs belting up and down the headlands and he's got plenty of his own woodland without me planting more trees. Then there's the farm staff to consider. I can't just go out and say thank you very much but I'm cutting down on farming.'

The take-up of set-aside in Europe in 1989 was equivalent to about 2% of the grain crop — an amount that could be a normal increase or decrease. The United States, which exports well over 100 million tons of food each year, has compulsorily set-aside 27% of its cultivable land area but is still looking to export surplus grain to Europe. A forecast by the Department of Land Economy at Cambridge suggests about 1.5 million hectares of land will be surplus to agricultural requirements by the year 2000. Is this really possible? Where does the figure come from? The General Agreement on Tariff and Trade war between the USA and Europe about exports/imports is entirely concerned with politics and economics, and does not reflect the world situation with regard to grain production. The European supluses in 1988-89 do not reflect the world situation in food supplies. Thus maize, wheat and soya stockpiles were 18% less than in 1984-85 and were below the Food and Agriculture Organisations guideline for world food security. The world population increase of 80 to 100 million per annum corresponds with an actual fall in total grain production and the forecast is that dire shortages could become common in the next century. We are indeed still clutching that maxim of looking after our own back-yard — what happens beyond is really no concern of ours. There does seem to be something morally wrong in hiding your talent, i.e. putting good land out of production, when there are billions of hungry mouths to feed. I don't know the answer, nor do I presume to point the finger, but somewhere, somehow there must be a better solution.

Doing nothing does not reflect the acumen of most farmers. This is why diversification has been preferred. The farmer's wife has been doing B and B for years to help pay for that week's holiday away from the farm, and the townies keep coming back. Diversification is an extension of this 'extra' — taking land out of food production and using it to provide a different source of income. A near relation to producing food is to produce crops for industry. Grain can be converted to ethanol for the motor car trade, and planted woodlands can be used for fuel and the building trade via coppicing, at the same time providing an amenity value.

The most popular diversion has been into the leisure industry, in particular theme parks and golf courses, the returns for which are high but require a good bank manager to provide the capital involved. Whilst theme parks are tending to reach saturation point, the demand for golf courses seems to be insatiable. When golf enthusiasts are to be found early in the morning sitting in their cars waiting for the course to open in order to get a game, there must be a shortage.

In the next century leisure time will increase and whilst holidays abroad will become ever more common, there will still be a substantial number of people who are attracted to their own environment — the countryside. The opportunity will be there for the farmer to select from the various ways to take advantage of this enormous market. Set-aside in its negative form may not be one of these ways.

8

FIFTY-FIVE QUOTATIONS AND COMMENTS

> 'A book that furnishes no quotation is *me judice* no book — it is a plaything.'
> (T.L. Peacock)

THE GREEN REVOLUTION

1. 'Green: is the signal to go ahead; has a strong Irish flavour; is nearly the name of a writer of fiction; is a sign of immaturity; is an indication of over-ripeness; is "home" to the golfer; are the eyes of the monster; is the colour of a garden pest; is the texture of the long awaited spring.'

(Author)

> Is it any wonder that, with such an admixture of similar and dissimilar ingredients, the Green Party started a revolution!

2. 'Any product you see on the shelf represents energy expended on its production or manufacture, packaging, transport and finally on its disposal.'

(Exeter Friends of the Earth 1989)

> How to become paranoic about the environment — follow the Green Code.

3. 'The blackboard and duster should be placed in one of the compartments of the box at the side of the blackboard ... this improves the cleanliness of the blackboard environment considerably.'

(Notice to Edinburgh Polytechnic Lecturers)

> Blackboards and dusters have been complaining for years but nobody took any notice until along comes a member of the Scottish Green Party whose favourite television programme is *Ever Decreasing Circles.*

4. 'Are the Greens really in the clouds, or at the cornerstones of our food production?'

(Mark Purdy, Green Party)

> The Greens have made a useful contribution to the farming world. It is when certain members suggest that the agricultural effort is misplaced that they are likely to be accused of living in cloud-cuckoo-land.

5. 'They [the people of India, Pakistan and Bangladesh] are being made to change their lifestyle so as to become totally dependent on technological gadgets and ... to throw their centuries old moral values overboard.'

(Al-Hafiz Masri, animal rights campaigner)

> There is nothing new in this. Some people call it progress. There will always be those who want to change to a different way of life, if the new way appears attractive. Moral considerations usually have to take a back seat.

6. 'I glanced at the tea-bag and thought that stuffed with chopped straw it could replace the annoying and impractical polystyrene chips.'

(Degenhard Urban, German entrepreneur)

> There is a wealth of possibilities for throw-away materials. Most of us are not simple-minded

enough to think of them.

7. 'Mr Smith says he used recycled toilet rolls. One can only marvel at the ingenuity of present-day recyclers.'
(H.J.S. Beasley, letter to *The Independent on Sunday*).

Mr Beasley has overlooked the confident assurance in Matthew 19:26, 'With men this is impossible, but with God all things are possible.'

FOOD AND THE FUTURE

8. 'Man has adapted food production . . . to provide for years of scarcity . . . enabling the possessor to employ his own labour and that of others in some other way, . . . advancement is always marked by a progressive increase in the quantity of non-food producing labour.'
(E.J. Payne 1892)

Shades of agrarian revolution, industrial revolution, strikes, food surpluses, set-aside, diversification and bankruptcies!

9. 'Green top may damage your health.'
(Department of Health warning)

That is to say, green top is on a par with cigarettes. Fresh milk described as 'raw' by its detractors is now taboo for schizoids who describe pasteurised milk as fresh and cannot distinguish good food from second-rate 'made-safe' offerings. A Health Minister who smokes is on a par with a Prison Governor who robs the inmates.

10. 'Food consumption is governed by the size of the human stomach.' (A deliberation from a 1957 government commission examining horticultural produce)

If only everyone had thought about this before we were landed with all those gargantuan surpluses!

11. 'Were another war to come, ploughs would tear up the structure-forming meadows and pastures, humus-providing animals would be slaughtered, soil improvement would be neglected, the vital reserve of soil fertility would be neglected.'

(G.V. Jacks and R.O. Whyte, *The Rape of the Earth* 1941)

Scientists may be worth listening to when they're talking about the present, but don't be fooled by their predictions for the future.

12. 'Our ability to model these changes [temperature] is not sufficiently advanced to say with any certainty exactly what the effects will be in the UK.'

(Dr Casson, National Environmental Research Council)

For a pundit to issue such reserved judgement is a pleasant change from the doomsday-style threats announced by less-informed public figures in high office.

13. Re the Greenhouse effect: 'The problem is that we are being asked to give answers with insufficient funding.'

(Dr M. Fasham, Institute of Oceanographic Science)

What you have to do, Dr Fasham, is not just fascinate your audience with pregnant possibilities, but put the fear of God into government at what would happen if you had to abort your project.

UNSAVOURY SALTS

14. 'The Department of the Environment has oversimplified the soil part of the nitrogen story to such an extent that it is coming up with the wrong answers.'

(Dr D. Powlson, Rothamsted Experimental Station)

Civil Servants have to simplify every bit of information in advising government ministers. The difficulty arises when both minister and servant forget to ask the scientist what the problem is about.

15. 'You won't find anyone who will tell you there is a scientific basis for the standard [50mg nitrate/litre of water].'

(R. Burgin, East Anglian local waterworks)

Conclusive proof that once you start a rumour, there is no telling where it will end.

16. 'We need to see if we can get more of this year's nitrogen into the crop and not into the soil organic pool.'

(R. Unwin, Senior Ministry of Agriculture Development and Advisory Service soil scientist)

The organic fraternity will be highly offended to hear that the disreputable stuff spread by inorganic cowboys has been invading their sacred organic precints and is actually being converted and integrating.

17. 'Where farmers are obliged to restrict their agricultural activities beyond the degree which could be regarded as good agricultural practice (G.A.P.), they should be compensated.'

(Ministry directive)

So speaks a cautious parliamentarian, who hasn't the faintest idea how to define good agricultural practice and who *suggests* compensation is *advisable* rather than a *necessity*.

18. 'Strict nitrate limits in the vulnerable zones of eastern and western England will reduce yield and competitiveness and will mean our food will cost more to produce.'

(Professor Greenwood, Agricultural Food and Research Council Wellsbourne)

May I remind Professor Greenwood of some fairly recent history when we were quite expert at importing cheap food. Now there will soon be the whole of Eastern Europe to tap and CAP notwithstanding, the weakest, or most expensive, will go to the wall.

19. 'There's no rationale for it [EC directive]. It cannot be done on a voluntary basis . . . Nobody is going to voluntarily bankrupt themselves.'

(Sir Simon Gourlay, National Farmers Union president)

There is no need for farmers to go bankrupt voluntarily. The government will do it for them after 1992.

20. 'We now have a number of drinking water supplies with unacceptably high levels of nitrate and pesticides and our rivers are too frequently at risk from leakages of farm waste.'

(Nicholas Ridley, speech to Agricultural Forum 1988)

How does someone, with a gift for bunging quite different dirty problems into a washing machine, qualify to assume responsibility for the environment?

THE NEGATIVE ELEMENT

21. 'In those countries which have achieved unparalleled advance in technological skill in medicine . . . we are witnessing the decay of man — his teeth, his arteries, his bowels and his joints on a colossal and unprecedented scale.'

(W.W. Yellowlees, conference at Royal College of General Practitioners 1978)

Fortunately or unfortunately his reproductive capacity may be bowed but remains unbroken. Since 1978 we are far more diet conscious, people live longer and are more active, smoking is taboo and sport and fitness have become a fetish — but there are still geriatrics around whose teeth, arteries, bowels, joints (and reproductive capacity) are not what they were!

22. 'In spite of overall improvement in the general health of the population there were definite hints of a deterioration in the group aged between fifteen and forty-four.'
(Sir Donald Acheson, report in *The Times* 1990)

This is the period of maximum sexual activity and tobacco usage — perhaps a combination of too much sex and too many fags. In the South West, I am sure the health problems are caused by too many Cornish pasties during the period of maximum intake!

23. 'Lepto [Leptospira hardjo] in cattle can cause abortion and milk drop and in man can produce symptoms similar to flu and meningitis.'
(*Farmers' Weekly* editorial 1988)

A control programme of vaccination is under way. How will the organic lobby meet the problem?

24. 'No biological control strategy could be used in isolation to control the cabbage fly.'
(Dr A. Thomson, Wellesbourne Vegetable Research Station)

Beetles are used in China, USSR and Denmark, but only subsidiary to pesticide control — a good argument for combined organic and inorganic treatments.

25. 'Never bury containers containing pesticides.'
(*British Agricultural Chemical Guide*)

They could have added cremation is cheaper, safer and less arduous!

VEGETARIANISM

26. 'Behold I have given you every herb bearing seed

which is upon the face of all the earth and every tree in which is the fruit . . . to you it shall be for meat.'

(Genesis 1:29, a text used by Joyce D'Silva, of Compassion for World Farming, in support of vegetarianism)

She has clearly not read Leviticus, where there are some forty verses in chapter 11 indicating which 'real' meats the Jews should eat.

27. 'It will be necessary to re-allocate the most productive land for the production of primary foods — that is to say, foods for direct human consumption without being processed through animals.'

(Compassion for World Farming)

This 'necessity' is based on the concepts of efficiency and economy of providing nutrition unqualified by any other consideration. If all human strategies were restricted by these limiting factors what a boring, tasteless, uneventful experience life would be.

28. 'How many gorillas are there left in the world?'

(Quiz question in the *Green Magazine*)

If the gorilla were the size of a flea nobody would be interested in the question, let alone the answer.

29. 'While I like the effect of sheep, I think that as animals they are very stupid.'

(B. Wookey, *'Rushall, the Story of an Organic Farm'*)

The apparent stupidity in animals is usually a reflection of the stupidity of the human in observation or behaviour.

30. 'But unlike dogs, cats tend to resist changes to their established diet, so making your cat a vegetarian may prove difficult.'

(Green Magazine)

This will be bad news for Jerrys! Incidentally I know quite a few humans who would be difficult to change to a vegetarian diet.

CARBON CONTROVERSY

31. 'Organic farming is a kind of agricultural apartheid separating the élite, Wordsworthian thinker from the low-caste Luddite labourer, and whilst providing only token monetary reward, qualifies for the exclusive Royal assent.'

(Author)

With apologies to all concerned.

32. 'Organic produce . . . is all wholesome food . . . its production does not damage their [the consumers'] environment or their health. Conventionally produced food cannot satisfy either of these desirable objectives.'

(D. Bull, letters in *Farmers' Weekly* 1989)

Such a belief needs no evidence. Mr Bull could have been a natural soul-mate of St Paul who is claimed to have said: 'Faith is the substance of things hoped for, the evidence of things unseen.'

33. ' So-called "natural" foods and remedies have no intrinsic advantage over "unnatural" ones and are often more dangerous.'

(W. Beswick, senior vice-president, British Veterinary Association 1990)

The life-blood of a controversy is the flow of pros and cons, but the flow is regularly diverted by unsubstantiated claims on both sides. Organic food may well be better than inorganic food. Nobody really knows (many do not need to!) 'Unnatural' remedies may be just as dangerous as 'natural' remedies. It all depends on the circumstances. Once we are rid of prejudice, then we can start to reason.

34. 'We are not saying it [organically produced food] is better than anybody else's or different except in terms of the way it has been produced.'

(Professor C.R.W. Spedding, chairman UKROFS)

> Euthanasia is another way of dying, but doesn't justify an elaborate intellectual organisation to identify rather than promote it.

35. 'We [the Soil Association] were unable, at that time, to show any significant superiority in organically grown crops.'

(Professor K. Mellanby, Talking Point, *Farmers' Weekly* 1989)

> Since nutrients absorbed by both organic and inorganic crops are essentially in an inorganic form, any *significant* differences are likely to be in cost of production. These will inevitably widen if oil prices go through the roof.

36. 'Organic farming does not cause any differences in flavour compared with conventional production methods.'

(D. Yourie, Brit. Soc. Anim. Prod. Conv. 1990)

> Another not particularly well-substantiated statement. Organic claims do tend to be more vociferous than inorganic claims, and so inorganic farmers must soldier on stoically, with increasing charges from the multi-nationals and increasing accusations from bewildered consumers.

37. 'By the year 2000 a fifth of British farmland could be converted to organic production — with public and political support.'

(A. Grant, chairman, Safeway)

> Wishful thinking perhaps by someone doing very nicely thank-you via customer gullibility.

38. 'Organic farming can harm health and welfare.'

(N. Carter, president, Veterinary Society 1990)
Naturally, Mr Carter rarely shops at Safeways.

39. 'Our standards [Soil Association] are drawn up with the help and advice of scientists, ADAS and veterinarians.'

> The change from top-gear production involving high inputs to specialised production involving low inputs has been an accommodation readily accepted by scientists, ADAS and veterinarians because they were the fountains of knowledge that led to all those surpluses. They can now bask in a blaze of glory as the wise men who stopped the rot before it became too cancerous.

40. 'I visualise . . . in a year or two's time, British [organic] produce . . . supplying an increasing share of the countries' supplies.'

(Richard Ryder, Parliamentary Secretary to Ministry of Agriculture Fisheries and Food 1989)

> Somebody should tell Mr Ryder that organic farming is an extensive system and cannot increase production by switching on the organic tap. The whole idea of 'organics', Mr Ryder, is to reduce the pressure on farmers, not encourage them to produce more to satisfy your export amibitions.

41. 'At present Tesco is importing 85-90% of its organically grown onions, 40% of the tomatoes, 60% maincrop potatoes, and 70-80% lettuce.'

(D. Wild, Tesco director 1990)

> No wonder the industry is screaming for more organic growers! Second-class citizens can only afford the inorganic stuff, so let's hope the swing's not too great!

42. 'The organic lobby has lost no time in assuring the public that none of this [listeria and salmonella contamination] would have happened if only the farmers of Britain

had not been seduced by the dreaded agro-chemical multi-nationals.'

(Oliver Walston, Talking Point, *Farmers' Weekly* 1989)

Well, of course, there is some truth in this organic assurance. The multi-nationals have provided the means to facilitate intensive farming systems, and close confinement of any kind does allow rapid proliferation of disease in circumstances where control programmes break down. Organic systems, however, could not provide equivalent supplies adequately or cheaply.

43. 'Why on earth didn't Prince Charles banish Oliver Walston to the Tower!'

(M. Purdy, letters to *Farmers' Weekly*)

This might well have happened when agro-chemicals were just clay-pipe dreams in an agrarian's head!

44. 'The most important cycle is the reconversion of dead organic matter into a form which will be taken up by the crop for its growth.'

(Vine and Bateman, *Organic Farming in Europe* 1981)

Dare one say this form is 'inorganic' or is this a taboo word in organic language?

45. 'Herbicides, pesticides and other technical chemicals are used to control the transfer of energy and nutrients as much as possible into the chosen agricultural products, such as wheat grain or lean meat, instead of allowing them to be distributed, as in the natural state, among other forms of life within the usual food web.'

(Vine and Bateman, as above)

This entry won the competition for the most intriguing sentence describing farming v wildlife. It needs to be re-read a number of times to appreciate the finer points.

46. 'Some of the practices of modern conventional farming have been to augment pest and disease problems and to cause environmental damage by pollution.'

(Lady Eve Balfour)

> The accusation is technically true, but what is significant is the extent of this effect. The benefits gained have undoubtedly been greater than the harm inflicted.

47. 'Organic farming is an attitude of mind, it is not a technique.'

(Lady Eve Balfour)

> But the thought must be father to the deed to produce anything at all.

48. 'We could not afford to sell our crops [organic vegetables] for any less. Half our crops were ruined in 1987 after a bad mildew and blight year. The large risk factor is always at the back of our minds.'

(G. Hughes, Norfolk grower 1989)

> With no vigorous programme for control the inorganic physical treadmill has been replaced by an organic mental treadmill.

49. 'Organic farmers appear to be a lot nearer to it [a perfect farming system] than those who have been led down the easy chemical road.'

(J. Bower, *The Farm and Food Society*)

> Beauty and perfection are in the eye of the beholder, Mr Bower. Those in the community who are short-sighted require the aid of inorganic spectacles. Some of these spectacles, however, are rose-tinted and the viewers are then able to magnify organic systems into Gardens of Eden.

50. 'Farmers must learn the priorities of the marketplace

and meet this [organic] market.'

(John Gummer, Minister of Agriculture, Fisheries and Food, 1990)

> There is something oddly Stalinistic about this directive. The marketplace, a conglomeration of misinformed consumers, would appear to be the fountain of knowledge, whatever the market demands — do it! No suggestion of educating the market, or educating the farmer to educate the market.

51. 'If artificial manures are no substitute for dung because they provide no humus, it is not surprising that we should turn to an older way of conserving fertility . . . the lea.'

(Michael Graham, *Soil and Science* 1940)

> There is rarely anything particularly new in farming. The practice of ley-farming was used by the Celts to allow land to recover from crop exhaustion. The old adage: 'To make a pasture breaks a man' would justify that qualifying Joadism: 'It all depends' on how you make the pasture.

PROPOSITIONS

1. Research and Development

52. 'The Government remains committed to agricultural research and development.'

(Baroness Trumpington)

> A politician's choice of words is often misinterpreted. What the above affirmation means is that there are far more important and pressing issues to support than agriculture to which (following the cuts) the government remains fully committed.

2. Set-Aside

53. 'The 40,000 hectares of land set aside could be an ideal entry into organic production — since it is effectively converted (or will be after three years).'

(Editorial, *Farmers' Weekly* 1990)

> The organic concept is developing rather like Aids — it's spreading into every quarter and there appears to be no cure for it. The multi-nationals are quaking because consumers and producers seem to be hell-bent on trying out this new experience!

54. 'If land [set-aside] is not managed for weed control and fertility building by using a suitable leguminous crop . . . there will be enormous problems in trying to farm it organically.'

(Bill Starling, chairman Organic Farmers 1990)

> The organic virus is unlikely to become established on farms which do not conform to good agricultural practice.

55. 'Set-aside illustrates how well its proponents are able to distinguish the subtle difference between surplus food in the West that nobody wants or can afford and food shortages in Third World countries from which people are dying.'

(Author)

> Is set-aside an embarrassing admission of failure?

9

CONCLUSIONS

'Life is the art of drawing conclusions from insufficient premises.'

(Samuel Butler)

'As often as a study is cultivated by narrow minds, they will draw from it narrow conclusions.'

(John Stuart Mill)

These quotations describe the sort of conclusions that every rational-minded person hopes to avoid. It is nevertheless virtually impossible to reach acceptable conclusions in a jigsaw of determinations where some of the pieces are missing. Once the critical events have passed, history will report whether these conclusions were right or wrong.

The most remarkable change in the eighties, certainly on the political scene, has been the emergence of the Green Party, nearly matching the change in attitudes which erupted within the Communist bloc. Whilst fun can be readily poked at some of the more extreme ideas and attitudes, the general feeling is that this party has awakened a consciousness in us all about our surroundings. There is still a long way to go — the inclination to throw litter around, common amongst those less well endowed between the ears, remains a very evident bad habit.

We are certainly more diet conscious, and the media is crammed with information about how to avoid those foods and habits which killed our fathers and grandfathers (females often escape, or rather delay, such fates). There is also a greater awareness of how many people in the world lack the basic needs we take for granted, but what is really disconcerting is that their numbers don't appear to get any smaller.

The importance of what we eat is matched by that of what we drink, that is the basic requirement, water, not the perennial curse, alcohol. People travelling abroad, especially when they're doing a Grand Tour of Third World countries, are far more conscious of the kind of drinking water available than when at home. The World Health Organisation, however, is looking to change all that — wherever we are, we shall probably be carrying our little gadgets to check the nitrate content of any water we are about to down. The nitrate level just might determine the length of our life-span, and the risk will remain up front until some new hazard is unearthed by over-active researchers.

Meat and eggs, part of the backbone of society's diet, have had a troubled passage during the eighties, exacerbated by a combination of avarice, ignorance and stupidity of the people behind the scenes. The propensity to panic is usually to be found in penny-wise people who lack the ability to assemble and analyse the relevant facts. Whatever the situation, the media are in a position to cash in, showing their expertise in sensationalism in order to feed their readers or viewers and incidentally line their pockets.

The inclination to vegetarianism amongst the younger generation is worth noting in terms of the effect on the next generation, always assuming that this new interest is maintained. Objections to meat-eating are held for a variety of reasons, some of which have been specified. The influence of Green Party policies, in this respect, on young minds together with increasing opposition to intensive farming systems will no doubt help to maintain this diet preference. Any general campaign to persuade the public that what they are eating is bad for their health and bad for animals cannot be recommended unless it is based on sound, rational and acceptable facts.

Wildlife, and in particular bird habitats, have been affected by agricultural intrusion in areas which are predominantly arable. When the initial decisions were made in the fifties and sixties to extend arable farming, there was no discussion as to any possible effects on wildlife — the current level of this awareness was barely surfacing at that time. This is not to say that farmers have not been aware of the wildlife environment — it is part of their background, almost taken for granted. They too have been aware of environmental changes in certain areas, but have attached less importance to them than people outside the industry not directly connected with the farming of land.

In the past, when biological and agricultural interests have collided, the biological interests have had to take a back seat where financial factors were critical. The current moves to encourage the establishment of wildlife habitats are developing in an atmosphere of reduced pressure to produce food together with some financial incentive.

The loss of certain species from some developed arable areas has, I feel, been over-dramatised by people who consider our environment should be fossilised. It is often forgotten that all types of animal have the ability to adapt to new conditions and it is an evolutionary truism that some will be more successful than others. Only the zoologist and the palaeontologist mourn the passing of the dodo.

The move to organics is a kick against the establishment — that is, against the fertiliser and agro-chemical companies that have been, and still are, a necessary part of the farming objects — to produce food and to earn a living. Some of the restrictions legislated by organic groups are difficult to accept as being rational, e.g. the use of artificial urea is banned although it has precisely the same composition as natural urea in urine. Veterinary restrictions too allow animals to move in or out of an organic system depending on whether a condition is reached whereby an animal has to be treated inorganically to prevent deterioration or death.

Attitudes, however, towards accepted practices and consciousness of our environment are steadily changing and where there is indication that new concepts are viable, the agricultural companies concerned should examine them thoroughly and adapt whenever and wherever necessary.

Studies of the environment indicate how long-term some

of the effects are likely to be. This does not devalue any programme of proceeding with caution where there is evidence that what mankind is doing to the environment is inexpedient rather than beneficial. There is some evidence that what we are doing may catch up on us, but such evidence is based on estimates, and estimates can be notoriously unreliable — a second ice-age was predicted some years ago, but more down-to-earth, weather forecasts, for only one week ahead, are rarely 100% accurate.

Caution was never the mother of invention, and the territory where angels fear to tread is where some of us dare to venture. Those who do are not necessarily fools, but careful observers may choose to regard them as such to defend their own hesitancy.

I am not being cautious in refusing to take sides in the carbon controversy. There is no virtue in saying whether organic or inorganic farming can solve our problems when the evidence is totally inadequate. What, however, is fairly clear, is that both systems have contributions to make. The points that I would make are that organic farming is limited by its own restrictions, by the amount of organic fertiliser available and by the degree to which it can expand given consumer demand. Inorganic farming, on the other hand, is likely to be restricted by increasing costs of inputs, by consumer resistance and by over-production.

Whilst preference for organic produce is supported by claims that it involves the use of less fossil fuel than that for inorganic produce, there has been no substance to the claim, i.e. a valid comparison has not been made. It has been established that more cultivations will be needed on arable organic farms to maintain weed control, i.e. using additional fuel. Any differencces which may be found in total fuel consumption between comparable organic and inorganic systems could well be of little significance.

Although claims are regularly made that the amount of cultivable land is decreasing as urban areas expand, there are probably still sufficient resources in the world to feed increasing populations, provided these increases are not exponential. It is the combination of politics, economics, distribution, religion and commitment which together frustrate any rational programme of putting food into mouths.

In the animal kingdom in general there is a commitment within the family or within defined social groups to care for its members. There is no consciousness among such groups of problems outside their particular local environment, and where interference is minimal and apart from natural catastrophes, such groups are successful. Humans are unique in being conscious of far wider strata of social groups and because compassion is a much stronger element in human make-up there is this feeling of responsibility for what goes on in the rest of the world. At the present time Action Aid is dealing with 17 countries with a total population of nearly 1,200 millions, many of whom, without aid, would not necessarily die, but would just exist rather than live. This kind of compassion can become a predominant driving force to the extent that the romance of missions in far-off lands brought into our sitting-rooms by television is far more attractive and seductive than the requirements of the down-and-outs two streets away.

Are we in fact our own worst enemies by having too much compassion for the poor and starving? Whatever is happening abroad, there is always the need to defend one's own patch, and the emergency of set-aside, a policy to produce nothing, may amount to just this kind of defence.

In the natural scheme of things populations would find their own levels subjected to a pendulum-swing between balance and imbalance in a context where the rich appear to get richer and the poor to get poorer for a given level of resources. It is convenient to blame the status quo, forgetting that our historical performance has brought us to it. It is even more convenient to blame the Creator for all the discrepancies. Once we are prepared to accept some of the blame ourselves, the process for reconstruction can begin.

What we can be fairly certain about is that fossil fuels will not last for ever and are certain to be increasingly expensive. We have turned from coal, oil and gas to nuclear power for a new source of energy and arguments for and against this preference continue to emerge. Is there no way that we can find or produce a source of fuel which will be sustainable and might serve to reduce, if not completely eliminate, the level of pollution? The answer is yes; why not grow it?

The use of cereals and other crops has already been adapted to produce industrial fuel. There is an unexplored wealth of plant material in the world which could help to solve this energy problem. An example which illustrates the kind of potential available is Amarantha or pigweed, a rather unpretentious plant, regarded as a nuisance, which can crop in a wide range of conditions to produce over ten tons per acre. The resulting crop can be converted into edible protein to replace the expensive animal protein in Third World countries. It is twice as efficient in converting solar energy as most other green plants but has barely been researched. It is not unreasonable to believe that there are other plants, unresearched, which could be utilised, perhaps using genetic techniques in development, in order to complement or eventually replace fossil fuels as a source of energy.

The extreme fragility of relationships, in a world which relies on fossil fuel as an energy source, needs no emphasis and it must be right to pursue even at some initial expense, a potential source which will ensure greater stability in both energy development and in these relationships.

A recent publication by British Gas on renewable energy cites six alternative sources together with their economic viability — not a mention of the possibility of growing this energy — and there are acres of land, set aside, waiting to be utilised. It has been estimated that there are less than fifty years' supply of oil reserves at current consumption rates. Is there not some urgency here, even allowing for undiscovered sources?

The world is replete with problems and it has been said that there is a solution to every one of them. The responsibility cannot be left to those bright people who periodically have brainwaves, or to adventurers and entrepreneurs who occasionally stumble upon solutions. What is needed is a concerted effort in research which is essentially a co-operative exercise requiring government financial support.

The options for one of these problems are probably growing all around us, and the fields are wide open.

POSTSCRIPT

The recent Earth summit in Rio de Janeiro has emphasised the need for a substantial decrease in the emission of carbon gases, particularly from countries with advanced technology, in order to decelerate the Greenhouse effect. The development of a less contaminating fuel, using plants, for vehicular transport could be a major factor in providing the means for this deceleration. On the European mainland, research is already established in growing rape for conversion to a diesel-type fuel. No such research has been considered necessary in the UK. A commitment to undertake more studies of such fuel alternatives should become a priority so that the twenty-first century will usher in at least one acceptable and sustainable energy source.

BIBLIOGRAPHY

(References in Chapter 8 'Quotes & Comments' not included here)

Action Aid — Correspondence
Agricultural Ecosystems and Environment (1983)
Agriland Ltd — Correspondence
Agscene May/June 1990
American Society of Agronomy (1984) *Organic Farming*
Balfour, E.C. *The Living Soil* 1945
Barraclough, D. and Jarvis, S. British Grassland Society Winter Meeting 1989
Bell, J.C., Palmer, S.R. & Payne, J.M. 1988 *The Zoonoses*
British Broadcasting Corporation *The Silent Revolution*
Bøckman, O.C. et al (1990) *Agriculture and Fertilisers*
Boering, R. (1980) Editor — *Alternative Methods of Agriculture* (The Dutch Report)
Bresson, J.M. and Vogtmann (1978) *Towards a Sustainable Agriculture*
British Farmer (1989/90)
British Organic Growers' Association Bristol — Correspondence
Cereals '89 Report — Royal Agricultural Association (1989)
Darwin, C. *Humus and the Earthworm* Reprint 1945
Dewhurst, M. Fertiliser and Manufacture Association — Correspondence
Dudley, N. (1986) *Nitrates in Food and Water*
Edens, T.C. et al (1985) *Sustainable Agriculture and Integrated Farming Systems*
Farmers' Weekly (1988/90)
Fertiliser Manufacturing Association — Information Unit
Garwood, E.A. & Ryder, J.C. (1984) *Nitrate Loss Through Leaching on Grassland*
Grier, B. and Vogtmann (1984) *Marketing and Pricing of Biological Products in West Germany*
Graham, M. *Soil and Sense* 1944
Grassland and Environmental Research Institute, Hurley — Correspondence
Green Magazine December 1990
Harvey, J. and Wilson, R. *Nitrates* — Farmers' Weekly Report 1988
Hill, S. & Ott, P. (1986) *Basic Techniques of Ecological Farming*
House of Commons Energy Committee Report — Chairman, Sir Ian Lloyd
Hydro Fertilisers — Correspondence
Jacks, G.V. & Whyte, R.O. *The Rape of the Earth* 1954
Jollans, J.L. (1985) *Fertilisers in UK Farming*
Kemira Fertilisers 1989 Correspondence
Kirchmann, H. (1985) *Losses, Plant Uptake and Utilisation of Manure, Nitrogen etc.*

Knorr, D. (1983) Editor *Sustainable Food Systems*
Lampkin, N. et al (1989) Collected papers on Organic Farming
Lampkin, N. *Organic Farming* 1990
Long, A. Vegetarian Society — Correspondence
Macduff, J.H. et al (1985) *Nitrate Leaching Losses from Grazed Grassland*
MAFF (1990) *Nitrate Sensitive Areas*
MAFF (1990) *BSE*
MAFF *Code of Safety on Pesticide Use*
MAFF Hatton Report
McCarrison, Sir R. *Nutrition and National Health* 1944
NRC Board of Agriculture (1989) *Alternative Agriculture* National Research Council
Norman, Felicity, Green Euro-Candidate Talking Point *Farmers' Weekly* 1989
Northbourne, Lord *Look to the Land* 1942
Pimental et al (1983) *Energy Efficiency of Farming Systems*
Porrit, J. Director Friends of the Earth, Wye College Conference 1988
Portsmouth, the Earl of, *Alternative to Death & Famine in England* 1944
Powlson, D. Rothamsted Experimental Station Arable Crop Research
Report and Recommendations of Organic Farming USA 1980
Royal Society Conference 1987 *Farm animals — It pays to be Humane*
Royal Society (1983) *The Nitrogen Cycle of the UK*
Royal Commission on Environmental Pollution Report 1989
RSPCA London — Correspondence
Ryder, R.D. *Animal Revolution*
Scofield, A.M. *Homeopathy in Agriculture* 1989
Spedding, Prof. C.R.W. (1983) *Biological Efficiency in Agriculture*
Stonehouse, B. (1981) *Biological Husbandry*
Stopes, C. & Woodward, L. (1988) *Consumer Demand and the Market for Organic Food*
Taylor, W.J.D. Cricket St Thomas Wildlife Park — Correspondence
Ulbricht, T.L.V. (1980) Editor Agriculture and Environment
Voisin, André, Fertiliser application 1964
Walston, O. Talking Point *Farmers' Weekly* 1989
Water Research Centre Prof. R. Packham Royal Society Conference 1989
Werner, P. & Von Alvensleven (1984) *Consumer attitude to Organic Food in German Federal Republic*
Woodward, L. (1981) Research projects in Biological Agriculture
Woodward, L. (1984) *The Alternative View. Organic Farming.*
Wookey, B. (1987) Rushall. *The Story of an Organic Farm.*